CELL
PHYSIOLOGY
AND
BIOCHEMISTRY

PRENTICE-HALL FOUNDATIONS
OF MODERN BIOLOGY SERIES

WILLIAM D. McELROY
AND CARL P. SWANSON, *editors*

BATES   *Man in Nature, 2nd edition*

BOLD   *The Plant Kingdom, 3rd edition*

BONNER AND MILLS   *Heredity, 2nd edition*

DETHIER AND STELLAR   *Animal Behavior, 3rd edition*

GALSTON   *The Life of the Green Plant, 2nd edition*

HANSON   *Animal Diversity, 2nd edition*

McELROY   *Cell Physiology and Biochemistry, 3rd edition*

SCHMIDT-NIELSEN   *Animal Physiology, 3rd edition*

SUSSMAN   *Growth and Development, 2nd edition*

SWANSON   *The Cell, 3rd edition*

WALLACE AND SRB   *Adaptation, 2nd edition*

WHITE   *Chemical Background for the Biological Sciences, 2nd edition*

CELL
PHYSIOLOGY
AND
BIOCHEMISTRY

3rd edition

WILLIAM D. McELROY
The Johns Hopkins University

PRENTICE-HALL, INC.

ENGLEWOOD CLIFFS, NEW JERSEY

TO ERIC
AND MARLENE

FOUNDATIONS OF MODERN BIOLOGY SERIES   WILLIAM D. McELROY
AND CARL P. SWANSON, *editors*

C—13-122176-0
P—13-122168-X
Library of Congress Catalog Card Number   73-140687

Current printing   10 9 8 7 6 5 4 3 2 1

PRENTICE-HALL INTERNATIONAL, INC., *London*
PRENTICE-HALL OF AUSTRALIA, PTY. LTD., *Sydney*
PRENTICE-HALL OF CANADA, LTD., *Toronto*
PRENTICE-HALL OF INDIA PRIVATE LIMITED, *New Delhi*
PRENTICE-HALL OF JAPAN, INC., *Tokyo*

THIS SERIES, FOUNDATIONS OF MODERN BIOLOGY, WHEN LAUNCHED A number of years ago, represented a significant departure in the organization of instructional materials in biology. The success of the series provides ample support for the belief, shared by its authors, editors, and publisher, that student needs for up-to-date, properly illustrated texts and teacher prerogatives in structuring a course can best be served by a group of small volumes so planned as to encompass those areas of study central to an understanding of the content, state, and direction of modern biology. The twelve volumes of the series still represent, in our view, a meaningful division of subject matter.

This edition thus continues to reflect the rapidly changing face of biology; and many of the consequent alterations have been suggested by the student and teacher users of the texts. To all who have shown interest and aided us we express thankful appreciation.

WILLIAM D. McELROY
CARL P. SWANSON

OUR DEEPENING UNDERSTANDING OF SOME OF THE BASIC SCIENTIFIC problems of biology, medicine, and agriculture has advanced with startling rapidity in recent years. A central factor in this advance is due, in major part, to the application of chemical and physical knowledge to cellular processes. Our basic understanding of the chemistry of the gene, the function of DNA in the coding for protein synthesis, the function of RNA in protein synthesis and possibly DNA synthesis, the mechanism of action of enzymes, vitamins, hormones, minerals, and various metabolic inhibitors has established the field of biochemistry or molecular biology in the forefront of the biological sciences. The fusion of molecular genetics and biochemistry into one field of "structure and function" has laid the groundwork and the basic theory for major advances in the biological sciences. And this is just a beginning, for our recent success encourages us to envision more significant advances in the immediate years ahead.

This book is a brief introduction to the basic principles of cell physiology and biochemistry. It is intended for the beginning student and because of this certain detailed information is purposely omitted. The specific facts selected, however, are designed to illustrate the fundamental processes underlying cell structure and function.

WILLIAM D. McELROY

CONTENTS

TO OBSERVE THE DRAMATIC PROPERTIES OF A SINGLE CELL, WE HAVE ONLY to peer through a small hand lens or a microscope. For years, biologists have marveled at these properties—at the shifting of the protoplasm inside the cell, at the changing size and shape of the cell, at the slow movements of cells away from or toward light. Below the reach of the microscope, however, lies a realm of smaller things, the world of the molecules and atoms that compose the ultimate structure of all matter. Biologists recognize the importance of the visible parts of the cell, but they are also aware that knowledge of the submicroscopic molecular pattern of the protoplasm is equally necessary to our comprehension of the structure and function of cells. In order to understand basic biological phenomena, therefore, the biologist today must be familiar with the latest developments in chemistry and physics.

The chief aims of the biochemist are to describe and analyze the chemical changes that occur in organisms. Investigations into the chemistry of living systems have shown that individual cells, whether of plants, animals, or microorganisms, are fundamentally similar in function despite vast differences in structure. Cell physiologists in turn are seeking to explain: the response of organisms and cells to their environment; the mechanism of cell growth, duplication, and reproduction; the ability of cells to take up nutrients from the environment; and the function and method of control of an organism's metabolic machine. With the discoveries made so far, biochemistry, a science that is still in its infancy, has greatly illuminated the functional aspects of organisms as they affect the fields of agriculture, medicine, nutrition, and other older disciplines. The effectiveness of biochemical knowledge in solving practical problems depends on our mastery of the fundamental principles of the chemistry of life. More and more young scientists in the discipline of biochemistry and other fields of science are turning their attention to some of the major societal problems. In the interdisciplinary approach to problems of air and water pollution, population growth, and the world food supply, biochemists have an important contribution to make.

Biochemistry, therefore, involves more than an unraveling by organic chemists of the structures of the working parts of cells. Biochemists and cell physiologists must make clear the relationship between structure and function in such a way that eventually we will be able to explain in chemical and physical terms many of the puzzling mysteries that still confront us: the nature of a cell's environment and its ability to resist and adapt to changes in that environment; the functional significance of the nucleus, of the mitochondria, with their array of enzymes, and of the microsomes and cell membranes that regulate the transport of materials into and out of the cell; the meaning of nutritional requirements and the importance of vitamins, inorganic ions, proteins, fats, and carbohydrates in cellular metabolism; and, finally, the ability of cells to utilize the stored or potential energy in chemicals and thus to enable organisms to move, speak, think, and reproduce.

These are some of the goals of biochemistry and cell physiology. Although our knowledge has grown at an unprecedented rate during the past 30 years, much interesting and vital work remains to be done. In this brief book, we shall introduce you to the essential aspects of biochemistry and cell physiology and discuss how they underlie the basic processes of life.

CELL STRUCTURE AND FUNCTION

ALTHOUGH CELL SIZE, STRUCTURE, AND FUNCTION ARE DISCUSSED IN detail in another book of this series,° it is necessary for us to review briefly the major cellular organelles and the role they play in cellular function. Although cells vary in size and form, most have a similar intracellular organization which includes a *nucleus* that is surrounded by a nuclear membrane (this structure is missing in bacteria and blue-green algae) and *cytoplasm* that is covered by a cell membrane (see Fig. 1-1). There are numerous smaller structures in the nucleus (nucleolus, chromatin, the hereditary material) and in the cytoplasm (mitochondria, endoplasmic recticulum, lysosomes, ribosomes) and we shall consider their importance to cell function throughout the book. Some of the types of cell used in cell physiology and biochemistry are shown in Table 1-1.

° See C. P. Swanson, *The Cell*, 3rd ed. (Englewood Cliffs, N. J.: Prentice-Hall, 1969).

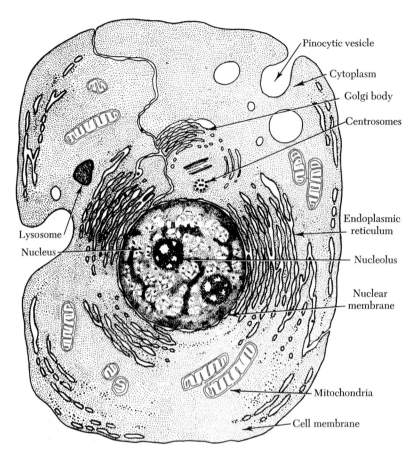

Figure 1-1 *Reconstructed cell, showing general cellular components as revealed by electron microscopy.* [Courtesy of Jean Brachet. Prepared by permission from an illustration copyrighted © 1961, 1963 by Scientific American, Inc. All rights reserved.]

THE CELL MEMBRANE AND THE OSMOTIC ENVIRONMENT

When a cell is placed in a suitable nutrient medium, it increases in size and divides into two daughter cells that are apparently similar to the mother cell. Just because a nutrient is close at hand outside an organism, however, does not mean that the organism is capable of utilizing it, because the substance must first pass through the cell membrane into the machinery that synthesizes cellular constituents. All the cell's food and waste products must pass through this membrane, and to do so a penetrating substance must be soluble to a certain degree in the fluid around the cell or in the protoplasm itself. Since not all dissolved substances can penetrate the membrane with equal facility, the membrane is said to be selective, and this selectivity is vital in maintaining the life of cells. Although the cell membrane is the primary determinant of the cell's internal environment, the other intracellular units—the nucleus, mitochondria, microsomes, etc.—have selective membranes that control their own inner environment. A number of factors, including hormones, ionic environment, metabolic energy, acidity, and temperature, affect the permeability of all these membranes.

Whenever two different solutions are separated by a selectively permeable membrane, an *osmotic system* is established. Each cell, therefore, represents an osmotic unit, since the selective membrane always intervenes between the inner protoplasm and the external solution, and large quantities of water and lesser amounts of dissolved substances constantly pass into and out of the cell across the membrane. Before we consider the complex process of osmosis, however, let us examine the spontaneous migration of molecules within the limits of a single solution.

If any dissolved substance is concentrated more heavily in one part of a solution than in another, it will spread gradually until its molecules

Table 1-1   Types of cells (modified after Giese)

NAME	TYPICAL SHAPE	EXAMPLES	PHENOMENA STUDIED BY EACH TYPE
Epithelial	Cuboidal or brick-like	Animal and plant epidermis	Permeability of cell membrane
Muscle	Spindle	Smooth and striated muscle cells	Irritability—conduction of electric impulse
Blood	Disc-shaped or ameboid	Red cells and leucocytes	Permeability, ameboid movement
Eggs	Usually spherical	Sea urchin eggs, frog eggs	Structure of protoplasm permeability, function of nucleus
Sperm	Usually flagellated	Sea urchin sperm	Motility, function of nucleoprotein
Protozoans	Diverse	Ameba, paramecium	Ameboid and ciliary movement, nuclear function
Unicellular algae	Diverse	Chlorella	Photosynthesis
Unicellular fungi	Diverse	Yeast	Fermentation and general metabolism
Filamentous fungi	Elongate	Neurospora, bread mold	Control of cell metabolism, biosynthetic pathways, general metabolism
Bacteria	Diverse	Colon bacillus	and nucleic acid function
Virus	Diverse	Tobacco mosaic, virus, bacteriophage	Function and structure of nucleoproteins

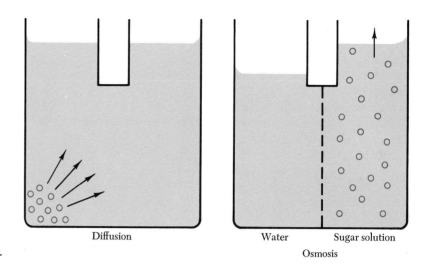

Diffusion		Water	Sugar solution

Figure 1-2   *Diffusion and osmosis.*

Osmosis

are evenly distributed throughout the whole solution. This process is called *diffusion*, and is caused by the random movements of all the molecules that make up the solution (Fig. 1-2). These movements are a manifestation of the kinetic energy of molecules, and the direction a chosen particle will take at any particular moment is entirely unpredictable, because its direction depends on its chance collision with other particles or with the wall of the container. Nevertheless, the *mass movement* of each kind of particle present in a solution can be predicted accurately on a statistical basis, for this movement is governed by the fundamental law of diffusion, which says that the particles of each different substance in a solution will diffuse from a region of great concentration to a region of less concentration. Diffusion will continue until every component reaches an equal concentration throughout the solution.

The relative concentration (the number of particles in a unit volume of solution) of each substance in solution is critical in determining the direction of the diffusion of that substance. Since no two molecules can occupy the same space at the same time, an increase in the concentration of any one substance necessarily displaces an equivalent amount of all other components in the solution. Accordingly, whenever the total solute concentration is high, the solvent concentration must be low, and whenever the concentration of one solute is increased, the concentration of any other solute and/or the solvent must undergo a corresponding decrease.

The velocity of diffusion is determined by a number of factors. Since the whole process depends on temperature, equilibrium is obtained more rapidly in warmer solutions, where the kinetic activity of the molecules is high. When the concentration difference is large, the particles have a greater tendency to escape from the concentrated region. Large particles diffuse more slowly than small ones, and the more viscous the medium the slower the diffusion. Equilibrium is reached very slowly when the distances involved are macroscopic, but within microscopic and ultramicroscopic limits, the concentration may become equalized almost instantaneously.

In biology, aqueous solutions predominate, and, consequently, osmosis may be defined as the exchange of water between the protoplasm and the solution that surrounds the cell. Take a simple osmotic system in which the solvent is water, which is separated by a selectively permeable membrane from a solution of sugar (see Fig. 1-2). In this case, both solute and solvent are under the same compulsion to diffuse, each away from the region of its own high concentration. In a perfect system, where the membrane prevents the sugar from moving, only the water is able to penetrate, and the two solutions can reach equilibrium only by the transfer of water from one solution to the other.

Since the water concentration on one side of the membrane is higher than on the other, water molecules will pass into the sugar solution and increase the volume of the water there until eventually the pressure of the column of water on the membrane compensates for the entrance of water. The pressure exerted by the solution is called the *osmotic pressure*. Earlier studies have indicated that a molar solution (one mole dissolved in a liter of solution) of sugar separated from water by such a membrane has an osmotic pressure of approximately 22.4 atmospheres at 0°C. This pressure can be related to the gas laws, since a gram molecular weight of a gas at atmospheric pressure occupies 22.4 liters, and if the gas is compressed to a volume of one liter it exerts a pressure of 22.4 atmospheres. We can thus calculate the osmotic pressure of a solution by using the following relationship: osmotic pressure $= CRT$, where C is the molar concentration, R is the gas constant (0.082), which we use to express the results in atmospheres, and T is the absolute temperature (273 + C°). Thus the osmotic pressure for a 1-molar solution of sugar at 0°C is: O.P. $= 1 \times 0.082 \times 273 = 22.4$ atmospheres.

Since the gas laws were developed for a perfect gas, they do not hold exactly for solutions, but they are quite satisfactory for calculating an *isosmotic solution* for cells, i.e., a solution that has the same osmotic pressure as the cellular contents.

This relationship between concentration and osmotic pressure is valid only for nonelectrolytes. For electrolytes (those compounds that can dissociate into two or more particles), the osmotic pressure is greater for a molar solution, since the pressure is determined by the number of particles. For example, if NaCl were 100 per cent dissociated in water, its osmotic pressure would be twice that of a molar solution of sucrose. In calculating the osmotic pressure of an electrolyte solution, therefore, we must multiply by the degree of dissociation.

The cell quite obviously is not a perfect osmotic system. Not only water but many dissolved substances commonly present in and around the protoplasm are able to penetrate the cell membrane in significant amounts, so that the exchange of water between the cell and its surroundings is accompanied by the exchange of other substances. Oxygen and carbon dioxide readily pass into and out of the cell. The permeability of the complex cell membrane depends not only on the nature of the surrounding particles, but also on the changing conditions inside and outside the cell.

Although permeability varies in different cells and sometimes on different sides of the same cell membrane, certain generalizations can be set forth. For example, we know that water penetrates most cells rapidly. Gases such as carbon dioxide, oxygen, and nitrogen, and fat-solvent compounds such as alcohol, ether, and chloroform easily penetrate all cell membranes. Somewhat slow to penetrate are such organic substances as glucose, amino acids, glycerol, fatty acids, etc., and slower still are the strong electrolytes—the inorganic salts, acids, and bases, and the large molecules of the disaccharides: sucrose, maltose, and lactose. Some cells can take up very large and complex compounds but almost none can absorb proteins, polysaccharides, or phospholipids.

There are many exceptions to these general statements. The bulk of all solutes present in protoplasm, including proteins and most sugars and inorganic salts, penetrate cell membrane very slowly if at all. The solvent water, on the other hand, enters and leaves a cell very quickly and this, coupled with the fact that water is more abundant than all other components combined, means that the water must bear the main burden of establishing an osmotic equilibrium between the cell and surrounding solutions. If a cell is placed in a solution with a water concentration drastically different from that in the protoplasm, so much water will enter or leave the cell that it may be destroyed. The cell wall in plants and in microorganisms, however, is usually rigid enough to prevent the swelling of the cell when water rushes in.

In an isosmotic solution, the concentration of water is the same as that in the protoplasm, and this condition occurs only when the total concentration of solute particles in the solution equals that in the protoplasm. This water balance is achieved because the water molecules that are continuously escaping from the cell are matched by an equal number entering the cell. A true isosmotic solution, therefore, contains a concentration of nonpenetrating or very slowly penetrating solute molecules that approximates the total concentration of nonpenetrating solutes in the protoplasm. An equal water concentration inside and outside the cell could not otherwise exist.

The best results in preparing an isosmotic solution are obtained by including salt mixtures in which sodium, potassium, calcium, and magnesium ions are represented in the proper proportions, since these ions are often essential to maintain the normal permeability of the cell membrane. A *hypotonic solution* contains a relatively low concentration of nonpenetrating solutes, compared to the protoplasm of the cell it surrounds, and thus the water concentration in it is relatively high. Cells placed in hypotonic solutions, then, tend to take in water and swell, in accordance with the fundamental laws of osmosis and diffusion. If the solute on the outside is very low, the swelling will continue until the cell membrane ruptures. When human red cells are placed in a solution containing only 0.2 per cent sodium chloride instead of the usual 0.9 per cent present in an isosmotic solution, the corpuscles swell and burst even before they can be observed under a microscope.

Cells placed in a *hypertonic solution*, on the other hand, tend to shrink, because the solution contains a higher concentration of solute molecules than does the protoplasm. In plant cells, the rigid wall maintains the original form of the cell, and the protoplasm draws away from the wall. Animal cells, however, when placed in a hypertonic solution shrivel up so much we cannot detect the original shape of the cell. When a red blood cell, for example, is placed in a concentrated sucrose solution, it shrinks from loss of water. If this process is not carried too far, it is reversible. Thus, if we place a plant cell in a hypertonic solution whose solute particles will slowly penetrate, the plant membrane will contract away from the cell wall (*plasmolysis*), but then as the solute particles slowly enter, water will return and the cell will resume its original size (*deplasmolysis*) (Fig. 1-3). Observing deplasmolysis, we can determine the rate of penetration of solute particles into plant cells, for the rate of deplasmolysis obviously parallels the rate of penetration by the solute particles.

If we remove the rigid cell wall from a bacterium or plant cell, we find that, like an animal cell, it is no longer able to maintain its shape. In a hypotonic solution, for example, the cell will burst. Normally, however, water enters the plant cells from a hypotonic solution until the protoplasm of the plant cell is forced outward against the

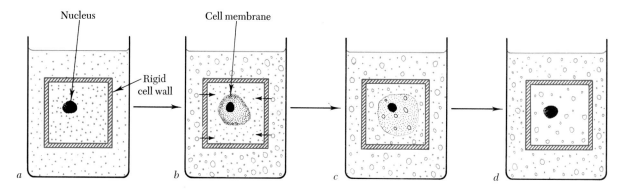

Figure 1-3   *Plasmolysis.* When a plant cell (a) is placed in a hypertonic solution of a slowly penetrating solute (b), water is rapidly lost from the cell, and the cell membrane shrinks away from the cell wall. As the solute slowly penetrates (c) and (d), water also re-enters, and the cell swells to resume its original size (d).

unyielding cell wall. When a sufficiently high internal pressure (*turgor pressure*) is generated, often amounting to several atmospheres, no more water can enter.

By placing the cells in varying concentrations of a sugar solution, we can estimate the number of osmotically active particles inside the cell. If we immerse a plant cell in a 0.5-molar solution of sugar and the cell neither shrinks nor swells, we know the cell contains the equivalent of 0.5 molar osmotically active particles (both electrolytes and nonelectrolytes), which would, in effect, exert a pressure of 12 atmospheres. Many plant cells, including bacteria, have this concentration of osmotically active particles inside. Cells such as bacteria that have their cell walls removed will thus swell and burst when placed in water, because they are unable to withstand the 12 atmospheres of pressure. Penicillin kills growing bacteria by preventing the synthesis of a new cell wall, thereby exposing them to the hypotonic solution that literally blows them up. To maintain these bacteria (called *protoplasts*) that have no cell wall, we increase the concentration of the solute—usually 0.4 to 0.5 molar sucrose.

ACTIVE TRANSPORT

We do not want to leave the impression that the selective membrane is passive and that simple diffusion and osmosis account for all exchanges between cells and the surrounding fluids, for many cells can accumulate certain substances and exclude others against the natural tides of diffusion. Many marine algae, for instance, accumulate iodine to a concentration which is more than a million times greater than that of the sea. The cytoplasm of cells generally is exceedingly rich in potassium and poor in sodium compared to the surrounding sap or lymph, and this situation cannot be accounted for by the simple laws of diffusion or osmosis. The cell, therefore, often spends energy to force the molecules of a particular substance to move against a concentration gradient (active transport). Since the energy to generate these forces is derived from the metabolic process, if metabolism is temporarily suspended, the cell loses its capacity to work against the tide, and the laws of simple diffusion and osmosis will control all exchanges. For identification, we customarily designate the moving machinery as "pumps," and thus speak of the sodium and the potassium pump.

We now know that there are specific structures in the cell membrane which regulate the pumping of substances into the cell. These units on the cell membrane are under genetic control and can be eliminated by a change in a specific gene. As we shall see later, a cell may contain an enzyme that catalyzes the metabolism of a particular substance but is unable to utilize the external supply of this substance because it cannot transport the substance through the cell membrane. The enzyme, therefore, can attack only those molecules that are produced internally. By gene mutation and selection, it is often possible to reverse this process and obtain a cell type that has a new specific pump. It is evident that the cell membrane is not a passive selectively permeable barrier, but rather an active partner of the metabolic machine, with many of the properties we normally ascribe to enzymes, the chemical catalysts of the cell.

One of the best ways to show this active pump is to use protoplasts of bacteria (bacteria with their cell walls removed). Certain strains of bacteria will accumulate the sugar lactose to such an extent that the protoplasts swell and burst due to the increased osmotic pressure internally. Since it is obvious in this case that the taking in of lactose occurs against a concentration gradient, it is not surprising that compounds (azide, cyanide) which inhibit cellular metabolism also inhibit the accumulation of the sugar.

There are many other examples of active transport, some of which will be considered in other books in this series. These examples include osmo-regulation (salt secretion) in fish and active re-absorption of salts and sugars by the tubules of the kidney° and the accumulation of salts by plant roots thus leading to a high internal osmotic pressure.† It is

° See salt secretion in Knut Schmidt-Nielsen, *Animal Physiology*, 3rd ed. (Englewood Cliffs, N. J.: Prentice-Hall, 1970).

†Arthur W. Galston, *The Life of the Green Plant*, 2nd ed. (Englewood Cliffs, N. J.: Prentice-Hall, 1964).

now clear that a large variety of biological processes are dependent on the active and selective intake of inorganic and organic compounds. Thus, the cell does not have to depend on the external concentration of a substance, for active transport makes it possible to regulate the entry of substances at a rate to satisfy the needs.

Unfortunately, we know very little about the actual mechanism of active transport. From what is known, it seems reasonably certain that the pumps in the membrane behave very much like enzymes, and consequently we must learn more about the mechanism of action of these biological catalysts.

Ameba and ameboid-type cells, such as white blood cells, epithelial cells of the intestine and others, can bring substances into the interior by forming invaginations of the cell membrane that finally pinch off and float free in the cytoplasm (see Fig. 1-4). These small vacuoles are called *pinosomes* and the general process is called *pinocytosis*. *Phagocytosis* is a similar process in which ameba and white blood cells are able to engulf large particles or bacteria by extending out around them "arms" (pseudopods) of the cell surface.

A number of different experiments indicate that only under certain conditions is pinocytosis observed. In water, for example, the addition of proteins or salts will induce the process, which will continue only for a limited time. There appear to be specific sites on the cell surface capable of forming the pinosomes and once these have been induced to do so by substances in the medium, the process of pinocytosis will continue only for a limited time. Restoration of the sites on the membrane is necessary before pinocytosis can begin again.

It is important to realize that the material in the pinosomes is not inside the cell in a metabolic sense. The membrane around the engulfed material persists and behaves very much like the cell membrane itself. Small molecules can readily diffuse into the cytoplasm but larger ones appear to be retained, and the vesicles decrease in size, gradually becoming part of the cytoplasmic granules.

The significance of pinocytosis in cell function is not entirely clear at the present time. It is one way of moving available macromolecules to the interior of cells and this may be important in certain cases where movement through the outer cell membrane would otherwise be impossible. It has been suggested too that the intake of the sperm by the fertilized egg is a process related to pinocytosis.

Figure 1-4 **Pinocytosis.** [Courtesy of Dr. David Prescott.]

2   THE
CHEMISTRY
OF CELL
CONTENTS:
INTERACTION
OF LIGHT
WITH
MOLECULES

USING SUNLIGHT AS A SOURCE OF ENERGY, GREEN PLANTS CONVERT, BY
the process known as *photosynthesis*, carbon dioxide and water into
all the organic molecules associated with life; the gas oxygen also
appears as one of the products. At the same time, the green plants
replenish the organic world, which is continuously being destroyed by
heterotrophic organisms. Life on this planet, therefore, is absolutely
dependent on light.

Photosynthesis is not the only biological process affected by light.
The vision of all animals is sparked by light energy striking specialized
cells in their eyes. The movement of plants and certain animals toward
or away from light and many other rhythmic biological phenomena
demonstrate that light energy is of overwhelming importance in the
functioning of organisms. Before discussing these photobiological
processes, we should consider the properties of light and its interaction
with molecules.

When a fine beam of sunlight is passed through a glass prism, it is resolved into its component colors: red, orange, yellow, green, blue, and violet. Isaac Newton demonstrated that if the light passes through a second prism that is reversed, the colored lights recombine to give white light; but if a single color is selected from the spectrum, no subsequent treatment can change it in any way. From this simple experiment, he concluded that white light is composed of many colors that can be separated into discrete units or particles. However, when evidence was presented which indicated that light rays could bend and spread, much as water waves fan out when a rock is dropped into a pond of water, this corpuscular or particle theory was rejected in favor of a wave theory. Since light always diffracts, so that the edge of a shadow is never perfectly sharp, the wave theory is consistent with many of the classical phenomena related to optics.

Numerous experiments, nevertheless, have revealed that light also resembles ordinary material particles. These "light" particles, which are designated as photons or light quanta, were discovered in early experiments concerned with the effect of light on solid matter. When light was beamed onto certain metal plates, electrons were ejected. The velocity of the ejected electrons was uninfluenced by the intensity of the light, although the higher the light intensity the greater were the number of electrons ejected. This is the basic principle underlying the photoelectric cell. When light hits a metal surface, the ejected electrons generate a small electrical current. Researchers directing colored light (i.e., light with a specific wavelength) onto a plate found that the velocity of an ejected electron was affected in such a way that the kinetic energy of the electron was inversely proportional to the wavelength or color.

The photoelectric effect led Albert Einstein to propose that light is in the form of small photons or quanta of energy. The energy of the photons determines the color of the light, and the entire energy of one photon is absorbed by an ejected electron, resulting in the destruction of the photon; the energy to eject the electron thus comes from that contained in the destroyed photon. Earlier, when Max Planck attempted to explain the energy distribution (in different wavelengths) of the light emitted by hot bodies, he discovered a fundamental constant that relates frequency to energy. This constant, h, is also the one needed to give the proper energies in the photoelectric effect. The energy, E, of a photon or light quanta could be calculated from the equation, $E = h\nu$, where ν is the frequency and h is Planck's constant.

Frequency has its usual meaning, normally expressed as the number of vibrations per second.

Consider the waves on the surface of a pond. When a rock is dropped into the water, a wave spreads out from the point of impact. If we drop rocks repeatedly, multiple waves are set up; the faster we drop the rocks, the closer are the peaks of the waves together. This distance between the peaks of two waves is the wavelength. Since the velocity of propagation of the wave remains constant, it is clear that the shorter the wavelength, the greater is the frequency. In other words, the speed of propagation divided by the wavelength equals the frequency. In the case of light, therefore, we can modify the equation describing the energy of a photon to this expression, $E = hc/\lambda$, where c is the velocity of light (3×10^{10} centimeters per second), and λ is the wavelength of light in centimeters. The actual value of h is 6.624×10^{-27} erg-seconds.

In biochemistry, we commonly talk in terms of calories per mole of a substance instead of in terms of ergs, and we use a variety of units for the wavelength of light. Radio waves are customarily measured in meters or centimeters, and visible light in either nanometers or Angstrom units. A meter (m) $= 10^2$ centimeters (cm) $= 10^3$ millimeters (mm) $= 10^6$ micrometers (mμ) $= 10^9$ nanometers (nm). The Angstrom unit (Å) is 10^{-8} centimeter and therefore equals 0.1 nm. Multiplying by the appropriate constants, we can write the above equation relating energy to the wavelength of light as:

$$E = \frac{2.86 \times 10^7 \text{ calories per mole}}{\text{wavelength in nanometers}}$$

A mole is equal to 6×10^{23} molecules of a substance, and the photochemical equivalent would be, therefore, 6×10^{23} photons. Thus, by making our energy calculations on this basis, we can speak of calories per mole as we do in ordinary chemical reactions. Blue light, with a wavelength of 450 nm, then, is equivalent to approximately 64 kilocalories per mole. We now see that different colors of light represent photons or quanta of different energies, those of blue light having much greater energy than those of red. We also note that the connecting link between the corpuscular and the wave theory of light lies in the fact that the energy of photons is directly proportional to the frequency of the light waves.

The word "light" is usually applied to what can be seen by the eye. Since scientists, however, speak of ultraviolet "light" and report that they can "see" other types of radiation by means of photocells and other cells, light to them means more than just the radiation that is visible to the unaided eye. The distinction between various types of light radiation is thus one of convenience; all radiations of the electromagnetic spectrum represent a single phenomenon, and their differences are only ones of wavelength and photon energy.

We know that when light quanta fall upon a metal plate, electrons are ejected. Light quanta also eject electrons from molecules in solution. In an atom or a molecule in a stable state, there are as many electrons around the positive nucleus or nuclei as the atomic number of the constituents. These electrons occupy the space surrounding the nuclei; some are close in and tightly bound by electrostatic forces and some are farther away and therefore more loosely bound. The momentum and spin of these electrons fix them in orbitals about the nucleus. Only a certain number of electrons can fit into each orbit, and the ones that are most weakly bound are in the outermost filled orbit. According to quantum mechanics, other orbits are "allowed" even farther out, but since it requires energy to move an electron away from its positive nucleus, the electron will tend to stay as close to the nucleus as it can. All chemistry occurs in these outermost or valence orbits. When two electrons are present as a lone pair in an outer orbit, the spins of these electrons must be in opposite directions, so we say that the electrons are "paired." (See Fig. 2-1.) If one of these electrons absorbs a light quantum, it can be boosted into a higher orbit, and the molecule is then said to be in an excited state.

The energy required to push an electron from one orbit out to another obviously depends on the energy difference between the two orbits, and we can determine this difference by shining light of various wavelengths onto the molecules. Since the energy in a photon of light can only be used on an "all or none" basis (that is, a *part* of a quantum cannot be utilized), only those photons that are effective in forcing the electrons to the outer orbits will be absorbed, and by investigating

Figure 2-1   Absorption of light quanta by molecules.

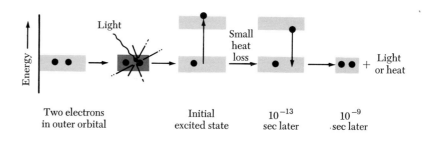

Energy →

Light

Small heat loss

Light or heat

Two electrons in outer orbital

Initial excited state

10^{-13} sec later

10^{-9} sec later

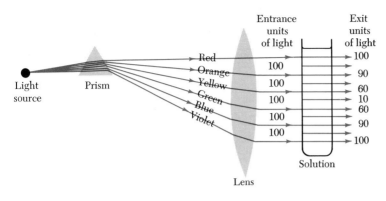

Entrance units of light 100 100 100 100 100 100

Exit units of light 100 90 60 10 60 90 100

Red
Orange
Yellow
Green
Blue
Violet

Light source

Prism

Lens

Solution

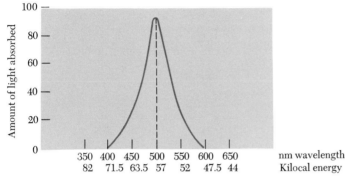

nm wavelength	350	400	450	500	550	600	650
Kilocal energy	82	71.5	63.5	57	52	47.5	44

Figure 2-2 **The absorption spectrum.**

which wavelengths of light are absorbed, we obtain what is called the *absorption spectrum* (Fig. 2-2). If we break up ordinary light through a prism, then beam the different colors through a solution of some chemical, and observe the light coming out the other side, we can determine qualitatively which wavelengths of light are absorbed by the molecules in solution. By using a photocell instead of our eye, we can quantitatively record the amount of light absorbed and plot this as shown in Fig. 2-2. Since the absorption peak is at 500 nm, this means that approximately 57 kilocalories of energy per mole, on the average, are necessary to push an electron from a filled to an unfilled orbit.

As we have indicated, when a quantum of light is absorbed by an atom or a molecule, an electron is boosted to a higher energy level. When the displaced electron returns to the original ground state, it can emit a quantum of light (the mechanism is called *photoluminescence*).

There are two fundamental types of photoluminescence: *fluorescence* and *phosphorescence*. Fluorescence is the emission of light from

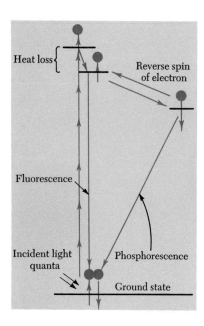

Figure 2-3   **Excitation of molecules** *to* *fluorescence or phosphorescence.*

substances only during the time they are exposed to radiation of various kinds, or for a very short time afterwards, whereas phosphorescence persists after the exciting radiation is cut off. In many cases, phosphorescence is at a longer wavelength (lower energy) than fluorescence, and this tells us something about the mechanism. As indicated in Fig. 2-3, in fluorescence the electron returns immediately to the ground state after excitation—by immediate, I mean in about 10^{-9} second. Usually this fluorescent light is at a longer wavelength than the exciting light, since some energy is lost as heat before the electron returns to the ground state.

In some cases, the electron may reverse its spin and move to an even lower energy state, but if it does it is almost impossible for it to return to the ground state until the spin is again reversed to the original direction, since no two electrons with the same spin can occupy the same orbit. The result is a long-term emission, a phosphorescence, of relatively low intensity that continues until something happens to reverse the spin. This emission may last for several seconds. We have described the fluorescent and phosphorescent states in some detail because these properties of molecules are becoming more and more important in the analysis of biological processes. When riboflavin phosphate is irradiated with ultraviolet light, for example, it emits a very brilliant fluorescence with a peak in the yellow-green range at 530 nm. When the flavin is reduced or when it combines with a protein, this fluorescence disappears or is greatly reduced, revealing valuable information about such reactions.

Although photoluminescence is the most common type of luminescence, other types are attracting the interest of scientists. *Electroluminescence*, for example, results when electrical energy excites molecules or atoms, and this type of cold light will probably provide much of the home lighting of the future. *Chemiluminescence*, a special concern of the biologist, is light emission caused by a chemical reaction.

MOLECULAR STRUCTURE AND THE
ROTATION OF LIGHT

As discussed previously, ordinary white light consists of many component beams of light, each of which has a characteristic wavelength. If we now consider just a single wavelength beam, say the Sodium D line, this light consists of waves vibrating in many planes at right angles to the direction of movement of the light beam. (Fig. 2-4.)

Certain crystals and other materials are available which permit only the passage of the light waves that are vibrating in a single plane. Such light is called *polarized* light, and the crystal or other material which selects for light vibrating only in a single plane is called a *polarizer*.

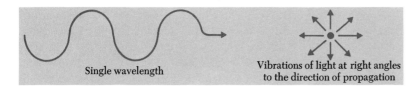

Single wavelength

Vibrations of light at right angles
to the direction of propagation

Figure 2-4

The plane in which the light will vibrate can be selected by simply rotating the polarizer. (Fig. 2-5.)

If the plane polarized light is passed through a second polarizer which is in the same orientation as the first polarizer, essentially all the light will be transmitted through the second polarizer. If the second polarizer is rotated 90° there will be no light transmitted. This phenomenon can be observed readily using the lenses of a pair of Polaroid sunglasses. If the two lenses are held a few inches apart, one behind the other, in front of a light source, and the lens nearer light is kept in a fixed position while the other is rotated, one can see the intensity of the transmitted light vary from very bright to very dim.

This is precisely the principle used in the instrument called the *polarimeter*, which is used to measure how much solutions of various compounds rotate plane polarized light.

It was observed many years ago that if solutions of certain amino acids or sugars were exposed to plane polarized light, the light which was transmitted through these solutions had been rotated into a plane

Figure 2-5

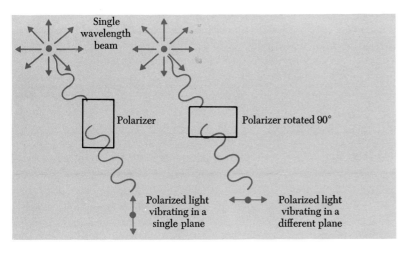

Single
wavelength
beam

Polarizer

Polarizer rotated 90°

Polarized light
vibrating in a
single plane

Polarized light
vibrating in a
different plane

THE CHEMISTRY OF CELL CONTENTS:
INTERACTION OF LIGHT
WITH MOLECULES

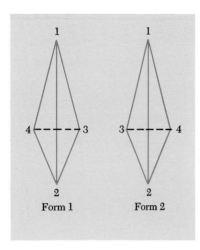

Figure 2-6

which was different from the incident light. Interestingly, in some cases the light was rotated to the right and in other cases it was rotated to the left. Such substances were said to be *optically active*.

Many years of work have led scientists to an understanding of why certain molecules rotate plane polarized light while others do not. In 1874 a theory was proposed independently by Van't Hoff and LeBel which explained optical activity. Essentially these workers stated that the four bonds of a carbon atom extend in different directions such that the molecule looks like a tetrahedron with the carbon in the center. If the four atoms or groups attached to the carbon are all different, then the molecule is asymmetric and can exist in two different forms (Fig. 2-6). This is easily seen if one uses toothpicks and different colored gumballs to construct the molecules shown in Fig. 2-6. These two forms cannot be superimposed on each other and therefore are different. Their relationship to each other is that of an object and its mirror image. Such compounds are *optical isomers* and are identical in all properties except their ability to rotate plane polarized light.

Any compound which contains a carbon atom with four different groups attached to it must be optically active and capable of existing in two forms. Common examples of optical isomers are found among the amino acids. The simplest amino acid, glycine, is shown below.

$$CH_3 - \overset{\displaystyle H}{\underset{\displaystyle NH_2}{C}} - COOH \qquad HOOC - \overset{\displaystyle H}{\underset{\displaystyle NH_2}{C}} - CH_3$$

This amino acid contains a carbon atom with four different groups attached to it and therefore is optically active. It can exist in the two forms shown. One of these forms rotates plane polarized light to the right and is called *dextrorotatory;* the other form rotates light to the left and is called *levorotatory.*

Since proteins are large molecules made up of many amino acids linked together, they are optically active. By measuring the rotation of light at many different wavelengths, one can obtain information about the structure of a protein even though it is very large. This technique of measuring rotation at different wavelengths is called *optical rotatory dispersion.* Although scientists understand the reason for the rotation of light by simple compounds, such as an amino acid, it is found that the rotation of light by a large molecule like a protein is more complicated than would be expected from the known properties of the amino acids. This is apparently due to some other structures in the protein. (See Chapter 3 for discussion of secondary structure of proteins.)

ALL BIOLOGICAL SYSTEMS, WHETHER OF MICROORGANISM OR OF MAN, require certain chemical substances and certain chemical changes in order to stay alive, grow, and reproduce. The fundamental unit of these systems, the cell, must be able to take up nutrients from the surrounding fluid, and must contain the machinery for creating new parts of itself from the food material. Some of the nutrients taken up become part of the structure of the cell, whereas others are broken down to provide the energy needed to synthesize new molecules. In general, all these nutrients are called food, and include carbohydrates, fats, proteins, minerals, vitamins, and water. Taken as a whole, these cellular processes are rather complicated, but if we examine the elaborate machine of the cell in the test tube, we can study the individual steps that lead to the synthesis of complex molecules.

During the past 30 years, biochemists have achieved remarkable success in isolating from cells or tissues specific parts of the metabolic machine. Once a particular chemical activity can be studied in the test tube, away from the complexities of the whole cell, rapid progress in the identification of the necessary factors involved usually results.

In such studies, the biochemist draws on all the knowledge available from the fields of pure chemistry and physics; indeed, biochemistry can be described as the blending of chemistry, physics, mathematics, and related subjects in an effort to explain the physiological aspects of cellular processes. As he has probed deeper into cellular function, the biochemist has discovered that certain groups of macromolecules are of unique importance to organisms. These are the proteins and nucleic acids. The latter we shall discuss in Chapter 10 when we consider the control of protein synthesis. At this time we must consider the proteins in some detail for it is in this group of chemicals that we find the *enzymes* which are the catalysts of chemical reactions in living cells. Enzymes are effective in very small amounts and are usually unchanged by the reaction they promote. The cell can duplicate itself in a few minutes or hours because its enzymes are capable of catalyzing the chemical changes associated with life processes. Since almost every individual chemical reaction in a cell is speeded up by a specific enzyme, we must know how these enzymes work.

The structure, function, and metabolic activity of a cell or tissue depend on the presence of specific protein molecules, for proteins play an essential role in all phases of the chemical and physical activities associated with life. The synthesis and function of enzymes (one class of proteins) determine the nature of cellular products. Enzymes are also intimately associated with such physiological processes as muscular contraction, nerve conduction, excretion, and absorption. Since the specific antibodies capable of combining with foreign or disease-producing agents are proteins, as are most of the important hormones that regulate cellular and tissue function, to analyze the physiology and biochemistry of cellular activity in some detail we must know as much as possible about these important macromolecules. We shall discuss first the structure of proteins and the methods of isolating and purifying them. In the next chapter, we shall investigate the nature of enzyme catalysis.

COMPOSITION OF PROTEINS

Proteins are complex substances of high molecular weight that contain nitrogen in addition to carbon, hydrogen, and oxygen. When proteins are broken down in the presence of water (that is, hydrolyzed) they yield a mixture of simple nitrogen-containing organic molecules called *amino acids*. The chemical composition of an amino acid is:

$$R-\underset{\underset{NH_2}{|}}{\overset{\overset{H}{|}}{C}}-C\underset{OH}{\overset{O}{\diagup}}$$

The —NH₂ group is the *amino group* while —COOH is called the *acid*, or *carboxyl*, *group*.° The *R group* can vary considerably; in fact, there are approximately 20 different R groups in nature and, therefore, 20 different amino acids. For instance, glutamic acid:

$$\underset{\underset{\displaystyle NH_2}{|}}{HOOC-CH_2CH_2\overset{\overset{\displaystyle H}{|}}{C}-COOH}$$

has an R group that is made up of two CH_2 groups and one carboxyl group. A few other amino acids are shown in Table 3-1. These are called *essential* amino acids because they cannot be manufactured by the body and therefore are indispensable in the diet of man. A person receiving a supply of nitrogen and carbon compounds as well as other nutrients can make the remaining *dispensable* amino acids.

In a protein molecule, the amino acids are joined together through bonds between the carboxyl group (COOH) of one amino acid and the amino group (NH₂) of another by the removal of water (Fig. 3-1). This bond between two amino acids (—CO—NH—) is called the *peptide bond*. If only two amino acids are linked together, it is called a di-peptide, if three a tripeptide, etc., and if a large number are connected,

° See E. White, *Chemical Background for the Biological Sciences*, 2nd ed. (Englewood Cliffs, N. J.: Prentice-Hall, 1970), for a general discussion of the chemistry of compounds of biological origin.

Table 3-1   **Essential amino acids**

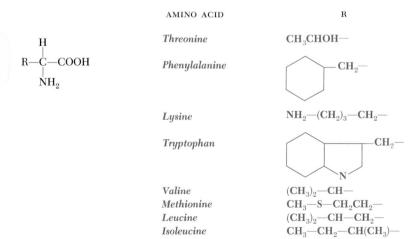

	AMINO ACID	R
$$\underset{\underset{\displaystyle NH_2}{|}}{R-\overset{\overset{\displaystyle H}{|}}{C}-COOH}$$	*Threonine*	CH_3CHOH-
	Phenylalanine	(ring)—CH_2-
	Lysine	$NH_2-(CH_2)_3-CH_2-$
	Tryptophan	(ring)—CH_2-
	Valine	$(CH_3)_2-CH-$
	Methionine	$CH_3-S-CH_2CH_2-$
	Leucine	$(CH_3)_2-CH-CH_2-$
	Isoleucine	$CH_3-CH_2-CH(CH_3)-$

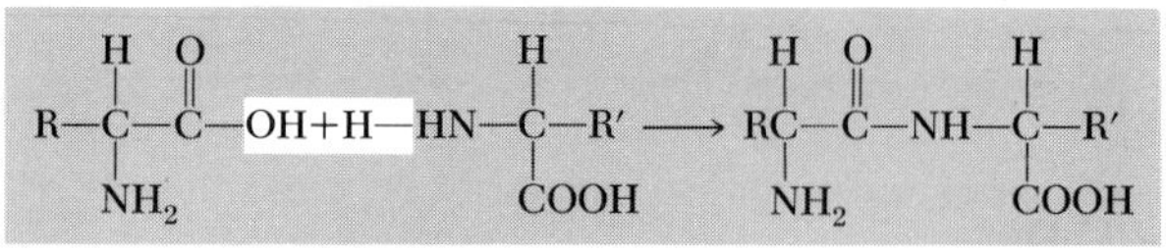

Figure 3-1   **Synthesis of a peptide bond.**

it is a polypeptide. The long polypeptide chain of a protein can fold in a number of ways to make unusual shapes or configurations, and in many cases additional bonds (cross links) help stabilize the folded structure. The folding of the long polypeptide chain is important in producing a stable and specific protein molecule, i.e., an enzyme.

The sulfur-containing amino acid cysteine is a key member of many of these cross linkages. As shown in Fig. 3-2 when two sulfhydryl (—SH) groups are near one another, they can be linked together by oxidation, i.e., by the loss of hydrogen (H), thus stiffening the folded polypeptide structure. The S—S link is called a *disulfide bond*, and it can be broken by substances which reduce it to the —SH form. The S—S linkage is also important in joining two or more polypeptides together. When we attempt to isolate and purify specific protein molecules from a cell, therefore, we should keep in mind that large aggregates or complexes of proteins can form by such cross linkages, in which case the apparent size of the protein molecules may be much larger than the basic unit we are interested in.

One of the important properties of proteins, its electric charge, depends on the acid and base properties of the individual amino acids. Consequently, before discussing this aspect of protein chemistry it is necessary for us to consider the nature of acids and bases.

Figure 3-2   **The formation of a disulfide bond** (cross link) in a polypeptide.

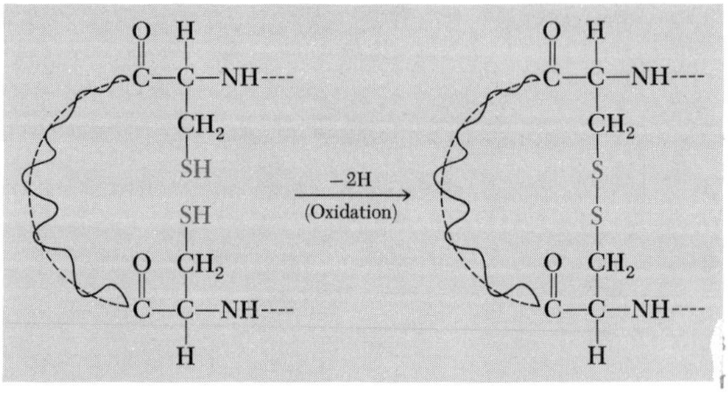

An *acid* is a substance that forms hydrogen ions (protons) in solution. The concentration of the hydrogen ions determines the degree of acidity of the solution. A *base* is any substance that combines with hydrogen ions. Thus, in an aqueous solution, hydrochloric acid dissociates into hydrogen ions and chloride ions (a base):

$$HCl \rightleftharpoons H^+ + Cl^-$$

NaOH, on the other hand, dissociates into the metallic ion (NA^+) and a base (OH^-). Consequently, NaOH can neutralize the acid by forming water ($H^+ + OH^- \rightleftharpoons H_2O$) and the salt, NaCl.

As indicated above, water dissociates into equal numbers of H^+ and OH^- ions and is therefore neither an acid nor a base. At ordinary temperatures, the actual concentration of H^+ (and of OH^-) in pure water is very small, about 10^{-7} moles per liter. A molar solution contains one mole of solute per liter of solution. For example, if 58.5 grams of NaCl (the gram-formula weight of Na, 22.991, plus that of Cl, 35.457) are dissolved in water and made up to a volume of 1 liter, we have a 1-molar solution. If 5.85 grams are dissolved and diluted to a volume of 1 liter, we have a 0.1-molar solution, and so on.

The concentration of H^+ times the concentration of OH^- is a constant and equals 10^{-14} ($10^{-7} \times 10^{-7}$). If we know the concentration of the one, then we can calculate the concentration of the other. Thus:

$$[H^+][OH^-] = 10^{-14}$$
$$\therefore [H^+] = \frac{10^{-14}}{[OH^-]}$$

If a 0.01 normal HCl solution, for instance, completely dissociates into H^+ and Cl^- ions, the $[H^+]$ will be $0.01 = 1 \times 10^{-2}$ gm moles per liter, and the $[OH^-]$ will be 10^{-12} gm moles per liter. It is convenient to express $[H^+]$ and $[OH^-]$ in powers of 10 and to make use of logarithms; thus, $0.1N = 10^{-1}N$, $0.01 = 10^{-2}$, $0.001 = 10^{-3}$, etc. The exponents -1, -2, -3, etc., are the logarithms of the concentrations. Instead of negative numbers, however, we can make them positive by taking the negative logarithms. A hydrogen ion concentration of 0.001, 10^{-3} gm moles per liter, therefore, has an acidity of 3, which is equal to $-(-3) = -\log [H^+]$. These numbers are called *pH values*, and they express the relative acidity of a solution; a pH value below 7 indicates an acid and one above 7 a base. Table 3-2 gives some examples of the relationship between $[H^+]$, $[OH^-]$, and pH. The pH of a solution or

Table 3-2

H^+ MOLES PER LITER	OH^- MOLES PER LITER	LOG $[H^+]$	$-$LOG $[H^+]$ = pH
1×10^0	1×10^{-14}	0	0
1×10^{-1}	1×10^{-13}	-1	1
1×10^{-4}	1×10^{-10}	-4	4
1×10^{-7}	1×10^{-7}	-7	7
1×10^{-10}	1×10^{-4}	-10	10
1×10^{-13}	1×10^{-1}	-13	13
1×10^{-14}	1×10^0	-14	14

the fluids of the body are kept constant by the operation of buffer systems. Most commonly, the buffer solution consists of a mixture of a weak acid and a strong base such as acetic acid and NaOH.°

ACID AND BASE PROPERTIES OF PROTEINS

Amino acids behave both as weak acids and weak bases, since they each contain at least one carboxyl group (—COOH) and at least one amino group (—NH$_2$). Such compounds are called *ampholytes.* An amino acid such as glycine, therefore, can carry a positive and/or a negative charge, depending on the pH of the solution (Fig. 3-3). The addition of hydrogen ions (H$^+$) to a solution of glycine suppresses the ionization of the carboxyl group, and the molecule acquires a net positive charge. On the other hand, adding a base (OH$^-$) removes a proton from the ammonium group, resulting in a net negative charge. At a certain pH, the molecule is electrically neutral (the number of + charges equals the number of − charges), and this pH, at which the dipolar ion will not migrate either to the positive or negative pole in an electric field, is called the *isoelectric point.* This point can be

° See E. White, *Chemical Background for the Biological Sciences,* 2nd ed., for a detailed discussion of buffers and pH.

Figure 3-3   *Amphoteric properties of amino acids.*

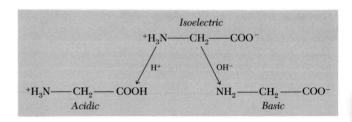

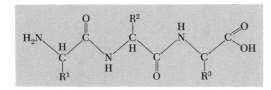

Figure 3-4   *A tripeptide.*

determined by titration with acid and base or by electrophoresis as discussed below. The pH of the isoelectric point depends on the dissociation constants of the basic and acidic groups.

Proteins behave as ampholytes because in the formation of a peptide bond the ampholytic character of the amino acids is preserved. A tripeptide is shown in Fig. 3-4. Note that on the right side we have a free acid group and on the left a free amino group; R^1, R^2, and R^3 refer to the remainder of the carbon skeletons of the three amino acids. The number of acidic and basic groups in proteins depends on the number and types of amino acids present. Obviously, if we linked a hundred glycine molecules together by peptide bond formation, we would end up with a large polypeptide with one free carboxyl and one free amino group.

There are amino acids, however, which have more than one basic or acidic group. Aspartic acid, for example, has two carboxyl groups, and, consequently, even when it is in the middle of a peptide where the terminal, or α, carboxyl is part of the peptide bond, another acidic group (the β carboxyl) is still free. Some amino acids, however, carry extra basic groups. Arginine, for example, has the following structure.

$$\boxed{H_2N-\underset{\underset{NH}{\|}}{C}-NH}-CH_2-CH_2-CH_2-\underset{\underset{H}{|}}{\overset{\overset{NH_2}{|}}{C}}-COOH$$

Arginine

The group shown in the square on the left called a guanido group is strongly basic. Another basic amino acid is lysine.

The behavior of proteins thus stems largely from the amino acid composition and the pH of the environment. What has been said about the charge properties of amino acids can be repeated for proteins, since they, too, have an isoelectric point and will migrate in an electric field. The pH relative to the isoelectric point determines whether they move to the positive or negative pole. The isoelectric point of proteins depends on the relative numbers of free carboxyl and free amino groups, which in turn depend on the amino acid composition.

To study the specific properties of enzymatic activities of a particular protein, we must first isolate it from other cellular components. Although often a difficult task, those who have patiently tried one procedure and then another and finally obtained a pure, homogeneous product have had the satisfaction of being the first to see a unique biochemical event. Certain general rules should be observed in isolating a particular protein. We need to know (1) something about the stability of the protein at different temperatures and pH and (2) in what solvents, other than water, the protein can be precipitated without destroying its properties. With these points in mind, we shall now discuss briefly the various procedures used to purify proteins.

EXTRACTION FROM THE CELL   We usually start by making a crude extract from cells or tissues that contain the desired protein, most often at a low temperature (2–5°C). The cells are broken open by a number of methods: by grinding, freezing, and thawing; by shaking with glass beads; by disintegrating them with sonic waves, etc. Water or a buffer is normally used to extract the protein; however, since the proteins are often associated with other cellular structures, such as fats, they may not be readily extracted with aqueous solvents. In some cases, the complex between fats and proteins can be broken by washing the tissue first with cold acetone or butyl alcohol.

ISOELECTRIC PRECIPITATION   Since the solubility of proteins varies with pH, by carefully varying the pH of the solution we can often precipitate the desired protein without destroying it—or we can at least eliminate other unwanted proteins. Even though the purification may not be perfect, this procedure is a useful one for concentrating the protein and at the same time eliminating a large number of compounds with low molecular weights. The precipitated protein can then be dissolved in a small amount of buffer.

SALT FRACTIONATION   High concentrations of salt are often effective in precipitating proteins from an aqueous solution. Some proteins are very soluble in high salt concentrations and others are readily precipitated, thus enabling a number of proteins to be separated. Ammonium sulfate has been the most useful compound for this salting-out purpose, and pH has also been an effective variable in this process. We employ salts and pH as variables for isolating proteins because they are multiply charged, and such charged groups can react with the water molecules or with each other. If proteins react with water molecules, their solubility is ordinarily increased. When ammonium sulfate reacts with the water, however, its effective concentration is lowered, and eventually the protein molecules will react with each other to form a precipitate (i.e., they become insoluble in water).

DIALYSIS   Some proteins (Euglobulins) are insoluble in pure water because the protein-protein interaction is stronger than the protein-water interaction. Such proteins can be brought into solution by the addition of a small amount of salt which interacts with the charge groups on the protein. The Euglobulins, then, can be separated from other proteins by lowering the salt content of the protein solution. This can be done by placing the protein solution inside a cellophane bag that is permeable to water and salt but is not permeable to the protein molecules. The bag is then placed in a container of water and the salt allowed to diffuse outward. This is dialysis (separation into two parts).

SOLVENT FRACTIONATION   Organic solvents such as acetone and ethanol are miscible with water and are therefore effective in the precipitation of proteins. Such solvent fractionations are usually carried out at temperatures below the freezing point of pure water.

In general, the methods discussed above are the most useful ones for the purification of reasonably large amounts of proteins; they are always able to keep the protein concentration at a high level, which is an important advantage, since proteins in dilute aqueous solution tend to unfold and denature. Additional methods have also been applied, and we will discuss them below when we consider purity.

SOLUBILITY   Since a homogeneous protein has a definite solubility in a given solution at constant pH, we can establish an accurate criterion of purity. If we add increasing amounts of protein to a solution, determine the amount that is dissolved, and graph the result, we should find that the solubility curve for a pure protein breaks sharply, as shown in curve A of Fig. 3-5. If the preparation contains two proteins, there should be two breaks in the curve. This technique for determining homogeneity is difficult to perform; in recent years other methods, depending primarily on electrical charge and molecular weight, are used more frequently.

ELECTROPHORESIS   Proteins, as we have seen, migrate in an electric field except at the isoelectric point, which is the pH where the net positive charge equals the net negative charge. Since a protein's net charge varies with pH, two proteins with different isoelectric points will obviously have different mobilities and so can be separated by the technique of electrophoresis. This technique is most effective when the protein solutions are buffered and at a known pH. Here is the way it works. For most proteins (whose isoelectric points are below 7), the

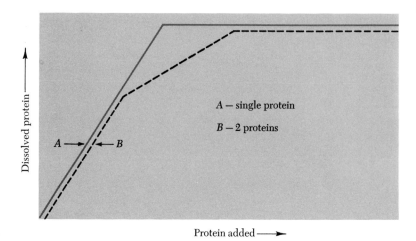

Figure 3-5 *Solubility curve of proteins.*

(on figure)

Dissolved protein →

A → ← B

A — single protein

B — 2 proteins

Protein added →

pH is made alkaline (7.5–8.0) to give them a maximum net negative charge. The solution is then placed in the bottom of a U-tube, and a buffer solution of the same pH is carefully placed on top of it (Fig. 3-6a). Electrodes are placed in the two arms of the tube, and the electric current is turned on. The negatively charged protein will move upward in the arm containing the positive electrode and downward in the arm containing the negative electrode (Fig. 3-6b).

We measure the rate of movement of the protein in the electric field by observing, through very sensitive optical instruments, the change in protein concentration at the boundary between the protein and the buffer (Fig. 3-6c). The change affects the optical properties of the area between the buffer and the protein solution; the incident light beam is bent the most as it passes through the solution in the region of the boundary, i.e., where the *protein gradient* is greatest. By using certain lenses, we can project what is called a *schlieren band* (Fig. 3-6e), the peak of which occurs at the boundary. The area under this curve is an expression of the concentration of protein. As shown in Fig. 3-6f, only one such peak appears with a pure protein. If two proteins are present, however, a discontinuity exists at the moving boundary, and two distinct patterns are formed (Fig. 3-6g). If these two proteins are similar in electric charge, just a shoulder may appear (Fig. 3-6h). A change in pH may change the mobilities of these two components, and better separation (and thus a better test for purity) can be achieved if the electrophoresis is carried out at a number of different pH values. Electrophoretic homogeneity, however, is not proof that the preparation is composed of a single molecular species.

Molecules of different molecular weight may have similar migration rates in an electric field.

A modification of this electrophoretic technique has proved even

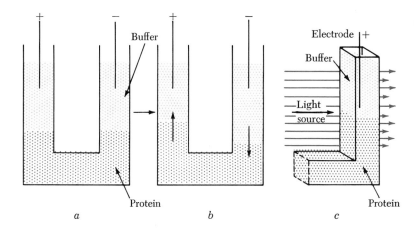

Figure 3-6   *Electrophoresis.*

a   b   c

Analysis of buffer-protein ascending (+) boundary

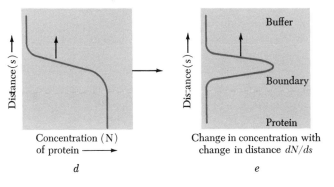

Concentration (N)
of protein ⟶

d

Change in concentration with
change in distance *dN/ds*

e

Electrophoretic patterns

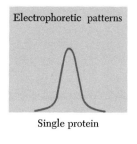

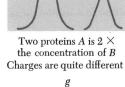

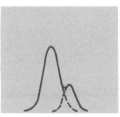

Single protein

f

Two proteins *A* is 2 ×
the concentration of *B*
Charges are quite different

g

Two proteins
Charges are similar

h

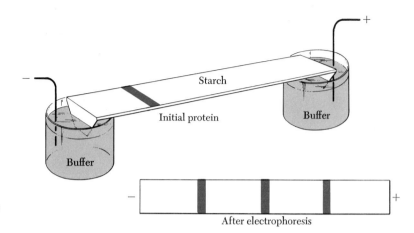

Figure 3-7 Electrophoresis using starch strips.

better in isolating proteins. A protein mixture is placed on a strip of agar, starch gel, sponge, or a piece of filter paper, which serves as a bridge between two tanks containing a buffer (Fig. 3-7). The movement of the proteins in the electric field is slowed by the protein-absorptive capacity of the supporting material. The advantage of this method is the relative ease with which protein components can be observed. With this sensitive technique, the heterogeneity of a number of highly purified preparations of proteins has been demonstrated.

SEDIMENTATION   The large size of protein molecules allows us to observe their movement in a centrifugal field. The rate of their movement, other things being equal, is directly proportional to their size, or molecular weight. In the centrifuge cell, the protein molecules move outward from the center of rotation, and a sharp boundary is formed between the pure solvent and the protein solution. By using the same optical method as in electrophoresis, we can study this boundary carefully. If one protein is present, we observe the same pattern as that at the bottom of Fig. 3-6. If there are two proteins, the heavier molecules will move ahead of the lighter ones. We must emphasize that apparent homogeneity in the experiment does not prove that the protein preparation is pure; under other conditions or by other procedures, heterogeneity may be observed.

As suggested above, the rate of sedimentation is an index to the molecular weight of macromolecules. Shape, however, as well as size influences the sedimentation rate, but for the purpose of calculating the molecular weight, the diffusion-constant of the protein molecules adequately takes into consideration the influence of shape. With these two constants, we can calculate the molecular weight.

SEPHADEX   Another efficient technique in the separation of protein molecules is the use of molecular series. A material called *Sephadex*, which consists of beads of an inert polysaccharide, is made in such a way that there are holes of uniform diameter in the granules. This material can be made with various size holes: very small, intermediate, or large. With the smallest holes, water is able to get in, but a large protein cannot. With intermediate size holes, water and proteins having molecular weights in the range of 12,000 to 50,000 can get in, but any protein larger than 50,000 cannot enter the holes. When Sephadex is used in a column it allows the separation of proteins on the basis of size. If we apply a solution of two proteins, one with molecular weight 10,000 and one with molecular weight 60,000 to a column of Sephadex with small holes or pore size, the smaller protein can get into the holes and will equilibrate with solvent surrounding the beads and solvent in the holes. Because the larger protein will not fit in the holes, it will move only in the solvent surrounding the beads. The fraction collected from the column will show the elution pattern seen in Fig. 3-8. Since the smaller molecule must pass in and out of the holes as it moves down the column, it will take much longer to elute. The larger molecule is just moving around the beads and is eluted very rapidly. Once a column is calibrated with proteins of known molecular weight, one can get an approximate molecular weight of an unknown protein by passing it through the same column and measuring carefully where it elutes.

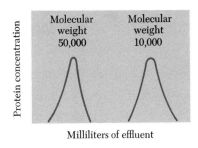

Figure 3-8   *Sephadex chromatography.*

Proteins are made up of a number of amino acids linked together in a definite sequence by a peptide bond. Some proteins may contain more than one peptide, and these are held together by specific cross links (i.e., disulfide bonds). The arrangement or sequence of the amino acids in the polypeptide is called the *primary structure* of the protein. In most proteins, the tightly coiled polypeptide chain produces a helical shape which we call the *secondary structure* of the protein molecule. Many of the interesting biological properties of proteins result from this secondary structure, and we shall consider it after discussing the amino acid composition.

AMINO ACID COMPOSITION   If proteins are treated with relatively strong acids (6N HCl) at 100°C for 10–20 hours, all the peptide bonds usually break and the amino acids are liberated. In recent years, a

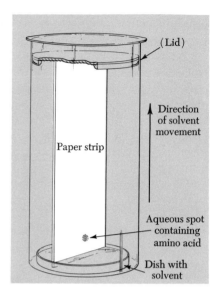

(Lid)

Direction
of solvent
movement

Paper strip

Aqueous spot
containing
amino acid

Dish with
solvent

Figure 3-9   Ascending paper chromatography.

number of methods have been developed for determining, qualitatively and quantitatively, the amino acid composition of such a mixture.

A method called *paper chromatography* is often used in the qualitative determination of the amino acid composition. The process is based on the property of amino acids separating on a strip of paper according to the relative solubilities of the acids in two different solvents. A small drop of an aqueous solution of an amino acid mixture is placed on a strip of paper which is placed in a closed box or cylinder so that the bottom of it dips into an organic solvent (alcohol, for instance) saturated with water (Fig. 3-9). The organic solvent moves into and up the paper, passing over the water spot containing the amino acids. Those amino acids that are highly soluble in the organic medium compared to their solubility in water will be carried along with the solvent, while those that are relatively more soluble in water will remain behind. Under constant conditions, each amino acid moves a specific distance up the paper. After the solvent has climbed a suitable way, the paper is removed, dried, and sprayed with ninhydrin, a chemical that reacts with each amino acid to produce a readily observable colored spot for each one. Since each amino acid moves a specific distance, by comparing a given spot with known amino acids, we can often make preliminary identifications.

When a large number of amino acids are present, it is often convenient to chromatograph first in one direction on the edge of a large sheet of paper, using organic solvent, and then, after drying the sheet, to rotate it 90° and rechromatograph, using a different organic solvent. The separation of a mixture of amino acids by this two-dimensional paper chromatography is shown in Fig. 3-10.

The principle of paper or partition chromatography has suggested other means of separating amino acids. For example, such substances as starch or silica gel can be substituted for the paper as a support for the stationary aqueous phase. The starch is combined with the organic solvent and water mixture and then packed in a column such as that shown in Fig. 3-11. The organic solvent, saturated with water, is added at the top of the column and allowed to flow until equilibrium between the water and organic phase is reached. A sample of the protein hydrolysate is then introduced at the top, and more solvent is allowed to pass through the column. The amino acids will move along the column at a rate that depends on their partition between the aqueous and the organic solvent phase. In addition, absorption of the individual amino acids on the starch is different and thus aids in their separation. The differences in the movements of the amino acids are often large enough to enable the acids to emerge from the bottom of the column completely separated from each other. By collecting small samples of the effluent solution, we can often make a complete quantitative separation of all the amino acids. In Fig. 3-11, four individual amino acids are separated on such a column.

The best procedure for separating amino acids quantitatively is one that depends more on the chemical properties of the amino acids. Since amino acids have ionizable groups with different dissociation constants, it is possible to separate them on columns that react with these charged groups. Ion exchange resins have proved extremely effective in this process and two general types, the cation exchangers and the anion exchangers, are used. When an amino acid mixture is added to such a column, the amino acids will exchange with one of the groups on the resin. The degree of exchange and the strength of the binding depend on a number of factors, including pH and the strength of the buffer. By continuously changing the buffer strength (gradient elution) and pH, we can quantitatively separate all the amino acids in a protein hydrolysate. With the aid of ingenious devices, we can continuously analyze the effluent from the column for its amino acid content. The results of a typical separation of a mixture of amino acids on a cation exchanger are shown in Fig. 3-12.

AMINO ACID SEQUENCE   When, by the analysis of a hydrolysate, we know the proportions of the amino acids in a protein, we can write a general empirical formula of the protein. But to describe the protein's structure more specifically, we must also know the actual sequence

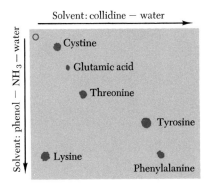

Figure 3-10 *Two-dimensional paper chromatography* of amino acids.

Figure 3-11 *Chromatographic separation of amino acids on a column.*

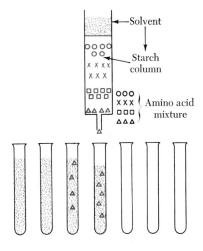

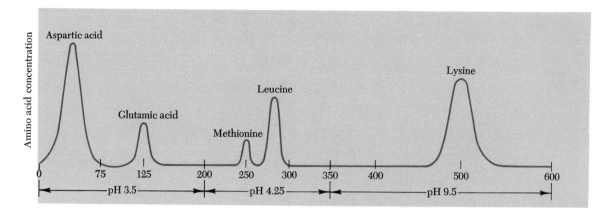

Figure 3-12   Separation of amino acids by ion exchange chromatography. The numbers along the base represent the volume of buffer collected from the column.

of the amino acids in the peptide chain, and whether it is a single or a multiple cross-linked polypeptide strand. The nature of the polypeptide strand is established by both physical and chemical methods. If the protein has only one polypeptide strand, it can have only one free α amino group (NH_2-terminal) and one free carboxyl (C-terminal) group. The free α amino group will react with reagents such as dinitrofluorbenzenes to give a dinitrophenyl (DNP) derivative. After such treatment, the protein can be hydrolyzed and the yellow-colored DNP-amino acid can be isolated by chromatography. If the protein is a pure one and we know its molecular weight accurately, we are in a position to say from the DNP analysis how many NH_2-terminal groups are present. A number of methods are available for determining the C-terminal amino acid residue. The enzyme carboxypeptidase is most helpful in this process, because it hydrolyzes peptide bonds that are adjacent to free α carboxyl groups. After treatment with the enzyme, the free amino acid can be analyzed quantitatively by chromatography.

To see if the protein contains more than one polypeptide chain, we split the cross linkages that hold the strands together and then observe whether there is a change in the protein's molecular weight. By unfolding the protein structure, we make the N- and C-terminal groups readily available for chemical reaction.

In Fig. 3-13, these procedures are briefly summarized. For purposes of illustration, we have used a protein that contains two polypeptide strands linked together by disulfide bonds. If we treat the protein with

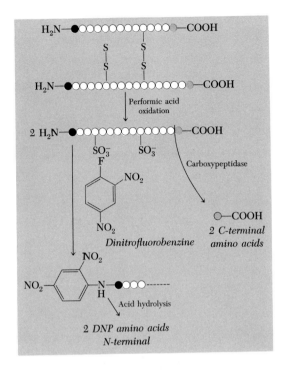

Figure 3-13   Determination of N- and C-terminal amino acid residues.

performic acid, the acid will break the bonds by oxidizing the sulfur, and, consequently, the molecular, or particle, weight will decrease by almost half. If the protein had been single-stranded, the molecule weight would have stayed essentially the same.

By following the above procedures, and applying other chemicals and proteinases such as trypsin and chymotrypsin, Sanger was able for the first time to determine the amino acid sequence in the protein hormone insulin. Insulin proved to contain two peptide chains, joined by disulfide bonds. In Fig. 3-14, the amino acid composition and sequence of beef insulin are shown. One chain (A) is composed of 21 amino acids, with glycine as the N-terminal amino acid and asparagine as the C-terminal. The second chain (B) is composed of 30 amino acids, with phenylalanine as the N-terminal and alanine as the C-terminal. The disulfide bridge in the A chain produces a ring structure that involves three cysteine molecules and alanine, serine, and valine. Insulin preparations from other sources differ from beef insulin in the amino acid arrangement of the cyclic structure of chain A. For example, horse insulin contains threonine, glycine, and isoleucine, while swine insulin contains threonine, serine, and leucine.

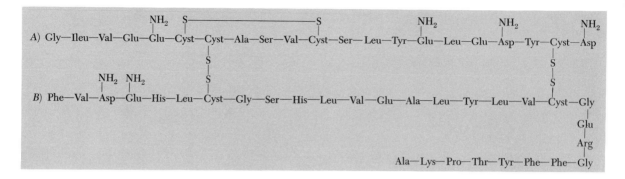

Figure 3-14   *Amino acid composition* and *sequence of beef insulin.*

Although the peptide bonds, the sequence of amino acids, and the disulfide cross linkages constitute the primary structure of all proteins, these organic chemical characteristics do not explain the behavior of most proteins in solution. Other characteristics, physiochemical in nature, must be attributed to the fact that the polypeptide chain is held in a coiled shape by forces other than those in the primary structure. Chief among these forces is the *hydrogen bond.* This bond results from the tendency of a hydrogen atom to share electrons with two other atoms, usually oxygen atoms. Take formic acid, for example. The early researches of Linus Pauling demonstrated that formic acid exists as a dimer because of hydrogen bonding (Fig. 3-15). The dotted bonds in the figure indicate that the hydrogen atoms share the electrons of the two oxygen atoms, each of which is in a different molecule, thus pulling the two molecules of formic acid closer together into a stable dimer.

Individually, hydrogen bonds are quite weak, but in a large molecule such as a protein, a number of them will reinforce one another and help stabilize the folded or helical structure. In the polypeptide chain, hydrogen bonds are mainly in the nitrogen and the double-bonded oxygen of the different peptide bonds (Fig. 3-15). In a typical helical protein, an α-helix, a complete turn contains 3.7 amino acid residues. Since each amino acid occupies the equivalent of 1.47

Figure 3-15   *Hydrogen bond formation.*

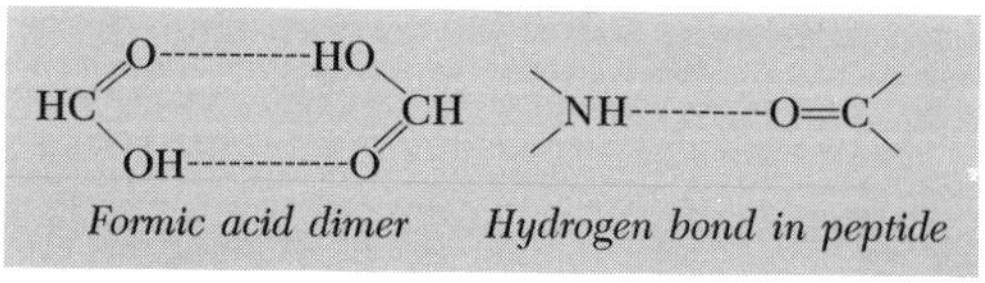

Formic acid dimer   *Hydrogen bond in peptide*

Angstroms in length along the central axis, every third amide group is linked along the chain by a hydrogen bond.

The helix is further strengthened by other types of noncovalent bonds. These interactions constitute the *tertiary structure* of the protein (Fig. 3-16). Considerable evidence now available supports the view that the secondary and tertiary structures of proteins are essential in biological function.

We have seen how vital water is in biological systems, and it is no less important in protein structure, since it is capable of participating in hydrogen bonds. Water apparently surrounds most protein molecules, through the formation of extensive hydrogen bonds with the protein. To function, proteins apparently must have an aqueous environment.

One final point may be made with regard to the tertiary structure of proteins. The availability of possible cross linkages between proteins will increase as the degree of coiling (and internal bonding) decreases. Therefore, fibrous proteins (long and relatively uncoiled) will tend to interact with one another through various weak linkages, and, as a result, solutions of proteins will tend to be more viscous than solutions of other substances. The cytoplasm of the cell is a viscous collodial system for this reason. Furthermore, the degree of interaction will be

Figure 3-16 **Some types of noncovalent bonds** that stabilize protein structure: (a) Electrostatic interaction; (b) hydrogen bonding between tyrosine residues and carboxylate groups on side chains; (c) interaction of nonpolar side chains caused by the mutual repulsion of solvent; (d) van der Waals interactions [Christian B. Anfinsen, The Molecular Basis of Evolution, Wiley, 1959.]

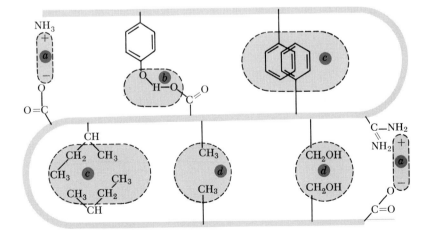

dependent upon the environmental conditions, since changes in pH, temperature, salt concentration, etc., will affect the nature or the availability of these cross-links. The observed changes in shape, viscosity, streaming, and so forth, characteristic of cells like the amoeba or cells in division, can be understood in terms of this changing collodial system, called a *thixotropic gel*. The study of the submicroscopic structure of cells has become one of the areas of active investigation during recent years. Although we cannot deal with this subject here, note that many properties of cytoplasm can be explained in terms of the secondary and tertiary structure of protein molecules.

4 THE CHEMISTRY OF CELL CONTENTS: ENZYMES

AN ENZYME IS A PROTEIN THAT CHANGES THE RATE OF A CHEMICAL reaction but does not affect the nature of the final products; in other words, it acts as a catalyst. It would be incorrect, however, to infer from this definition that the enzyme does not participate in the reaction.

Catalysts are characterized by the following properties. (1) They are effective in very small amounts. When analyzing the catalytic power of an enzyme, we should know the amount of starting material, or substrate, that is converted to product in a unit time by a given quantity of the enzyme. This quantity is called the *turnover number* of the enzyme and is defined as the number of moles of substrate converted into product per minute by one mole of enzyme. The turnover numbers of enzymes vary greatly, ranging from one hundred to over three million. (2) Catalysts are usually unchanged in the reaction. This property of ideal catalysis can only be approximated by enzymes, however, since they are not completely stable under most conditions. Proteins are extremely unstable substances and are easily inactivated or denatured by high temperatures or very alkaline or acidic conditions.

An extreme example of denaturation occurs when an egg white is boiled to harden it. (3) Catalysts ordinarily have no effect on the equilibrium of a reversible chemical reaction; they merely speed up the reaction until it reaches equilibrium. The function of the true catalyst, therefore, is to hasten the process in either direction. (4) Catalysts exhibit specificity in their ability to accelerate chemical reactions; that is, a given catalyst affects only certain types of chemical reactions. We shall speak of this specificity at greater length later.

One of the most dramatic advances in biochemistry during the last 30 years has been the isolation of a variety of enzymes from different cells, thus enabling us to study their mechanism of action in solution away from the complications of cellular activity. The complicated reactions involved in carbohydrate, fatty acid, and amino acid metabolism have been analyzed step by step. The process of isolating enzymes from cellular extracts and purifying them has been a slow but a rewarding one.

On the basis of an impressive body of information, we now believe that enzymes are definitely proteins. To understand how enzymes exert their catalytic action, therefore, we must know the details of protein structure. Since large gaps still remain in our knowledge of this structure, many aspects of the mode of action of enzymes are still obscure, but this does not bring us to a halt, for the properties of enzymes as catalysts may still be studied without immediate regard to the mechanism of action. From such investigations, valuable data about how enzymes act in a biological system have been obtained.

NAMING ENZYMES

Naming enzymes is very easy once we know the type of reaction or the substance (substrate) acted on by the enzyme. Enzymes are denoted by the suffix -ase. For example, an enzyme that catalyzes the breakdown of proteins is called a *proteinase*, one that catalyzes an oxidation is called an *oxidase*, and so on.

EFFECT OF TEMPERATURE ON REACTION RATE

As we have said, the function of an enzyme is to increase the rate of a chemical reaction. In water, the molecules of a cell are in ceaseless thermal motion, and occasionally react when they collide with one another. In a solution without enzymes, the chance that a molecular collision will result in reaction is very small; but if the appropriate enzyme is present, the probability will be greatly increased. The major question is: How do enzymes perform this unusual catalytic activity?

The answer seems to lie in the ability of enzymes to make a substrate molecule more labile, that is, more reactive to other molecules in the environment.

Chemical compounds that can be isolated are more or less stable, for unstable molecules, by definition, react with other molecules in the environment until they form more stable products. Stable molecules can react rapidly, however, if they are activated by the addition of energy. Arrhenius was the first to point out that all the molecules in a given population do not have the same kinetic energy. Some molecules, through collisions, acquire more energy than others, and these energy-rich molecules are more likely to react than energy-poor ones. In other words, there is an energy barrier to reaction, and the higher the energy barrier for a molecule, the greater is its stability. The energy required to hurdle molecules over the barrier is called the *energy of activation*.

Figure 4-1 charts a hypothetical reaction in which A is converted into B. This diagram holds for all chemical reactions, although the height of the energy barrier varies from one reaction to another. Note that for A to be converted into B, it must first acquire the necessary energy to form the activated molecule $A^{\ddagger}$. The inherent rate of a chemical reaction depends on the number of $A^{\ddagger}$ molecules existing at any one moment and the frequency of their decomposition (C) into products, B; or:

$$\text{rate} = (A^{\ddagger})C$$

It turns out that C is a constant and is essentially the same for all chemical reactions. In order to describe the absolute rate of a chemical reaction, we must find a way to determine the concentration of $A^{\ddagger}$. It can be shown that the number of activated molecules is proportional to the energy of activation, E, and consequently the rate of the reaction can be written as: rate $Ce^{-E/RT}$, where $e^{-E/RT}$ is proportional to the number of molecules in the excited state. R is a constant and T is the temperature. We determine E by studying the rate of reaction at different temperatures. When we plot the logarithm of the rate against the reciprocal of the absolute temperature, we obtain a straight line whose slope equals E/R, i.e., $\ln \text{Rate} = \ln C - E/R \times 1/T$ (see Fig. 4-2). In certain texts $\Delta H^{\ddagger}$ is used for the experimental energy of activation instead of E.

It is evident that as the temperature is raised, the rate of the reaction increases logarithmically. As a general rule, for every $10°C$ rise in temperature, the rate of chemical reactions increases 2–3 times (Fig. 4-3). The rise in temperature increases the number of activated molecules by increasing their movement and the number of collisions.

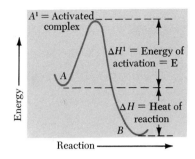

Figure 4-1   Energy of activation for a chemical reaction.

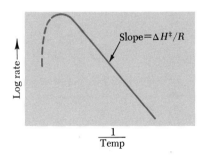

Figure 4-2   The experimental energy of activation ($\Delta H\ddagger = E$) is determined by plotting the logarithm of the rate at different temperatures against the reciprocal of the temperature.

THE CHEMISTRY OF CELL CONTENTS:
ENZYMES

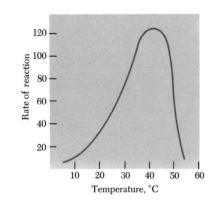

Figure 4-3 ***The effect of temperature*** *on an enzyme-catalyzed reaction.*

The rotation and vibration of molecules also increase with rising temperatures, and all these processes produce a greater proportion of molecules that have the necessary energy for reaction.

What has all this to do with enzyme-catalyzed reactions? One of the significant features of enzymes is their ability to lower the energy of activation, enabling reactions to occur at ordinary physiological temperatures. Exactly how they do this we do not know. But thanks to them, reactions that might ordinarily occur only at boiling temperatures, because of the energy barrier, can take place with relative ease in the cell.

No one has yet succeeded in isolating an activated complex. Activated complexes have only been deduced from kinetic studies of enzyme-catalyzed reactions carried out at different temperatures. Temperature affects ordinary chemical reactions in the same way it affects enzyme-catalyzed reactions. At high temperatures, however, the heat tends to destroy (denature) enzymes so that no further reaction can take place. Biological phenomena thus have optimal temperatures at which they function. Above a critical temperature enzymes do not perform. As shown in Fig. 4-3 for a hypothetical situation, the reaction rate increases rather rapidly and logarithmically from 0° up to about 25°C, but above this point, the increase slows, and at approximately 35°C the rate starts to decrease. If the damaging temperature is maintained for too many minutes, the enzyme will be completely inactivated.

The thermal behavior of enzymes imposes serious limitations on organisms; most cells lose their capacity to carry on metabolism at temperatures above 40°C. Only the few organisms that have heat-resistant enzymes are able to survive in exceptionally hot places. In most cells, the rate of metabolism, and hence the intensity of the life processes, changes as the temperature varies from day to day and from season to season. The winter metabolism of most organisms subsides to a point where dormancy is inevitable. Warm-blooded organisms such as man and a few other vertebrates are the only creatures that have evolved a method of controlling their body temperature.

ENZYME-SUBSTRATE COMPLEX

As is probably evident from our previous discussions, enzymes do more than lower the energy of activation. They seem, in some cases, to set the pathway or direction in which a particular substrate will be metabolized. For metabolic processes to occur, the substrate molecule must come into intimate contact with a specific enzyme. By combining with the enzyme, a deformation in some of the bonds of the substrate

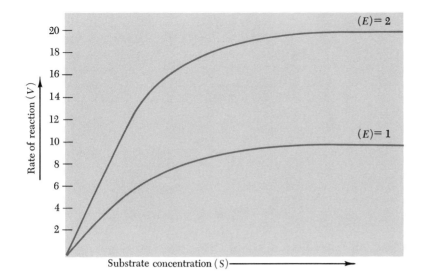

Figure 4-4   **Effect of substrate concentration** on the rate of enzyme-catalyzed
reactions.

molecule probably occurs, thus producing a condition that favors the reaction.

The idea that the substrate must combine with the enzyme before catalysis can proceed is an old one, and there is much experimental evidence to support it. If the substrate molecules must combine with the enzyme before catalysis takes place, collision, then, must play an important part in the process. Since the enzyme molecule is very large and the substrate molecule is usually quite small, for the substrate to combine with the enzyme there must be a relatively large number of substrate molecules in order to increase the probability of correct collision.

When we study the rate of an enzyme-catalyzed reaction involving different amounts of substrate molecules, we observe that the reaction rate does indeed vary with the concentration of substrate, at least over a limited range of substrate concentration (Fig. 4-4). The rate curve at first rises, then slopes off, and finally reaches a constant maximum value. Where the rate of the reaction no longer changes with an increase in the substrate concentration, we assume that the enzyme surface is completely covered or saturated with substrate. Two different enzyme concentrations are shown in Fig. 4-4. Note that the maximum rate (V_m) depends on the enzyme concentration.

The combination of a substrate with an enzyme is a very specific process, i.e., each catalyst has a unique surface containing a precise spot where the substrate molecule can join it. Because of the unusual

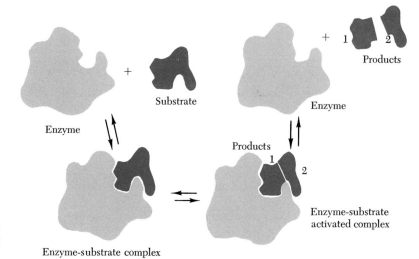

Figure 4-5   The formation of an enzyme-substrate complex, followed by catalysis.

folded surface of proteins, only very specifically shaped molecules can gain access to the particular chemical groups (such as SH). For this reason, the enzyme-substrate complex is often compared to a lock and key, as shown in Fig. 4-5, where the substrate is analogous to the key. We can write the reaction which describes Figs. 4-4 and 4-5 in the following way:

$$E + S \underset{k_2}{\overset{k_1}{\rightleftharpoons}} ES \overset{k_3}{\longrightarrow} E + \text{products}$$

Here k_1 refers to the rate of combination of E and S to form ES, and k_2 is the rate for the dissociation of ES to E and S. After the substrate combines with the enzyme to form ES, we assume that enough activation energy is acquired so that ES is converted into the product of the reaction, which dissociates from the enzyme surface to allow the enzyme to combine with another substrate molecule. The rate of conversion of ES to the products is indicated by the constant k_3. At a low substrate concentration some of the enzyme is free, and the maximum rate is not observed. With excess substrate, all the enzyme is converted into ES, and the reaction takes place at its maximum rate—i.e., maximum velocity $(V_m) = k_3(ES) = k_3(E)$.

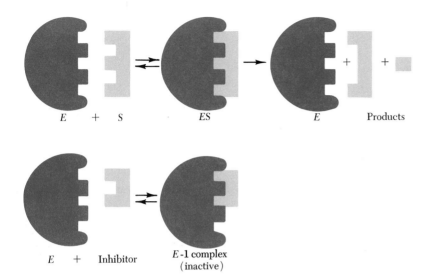

$$E \quad + \quad S \qquad\qquad ES \qquad\qquad E \qquad\qquad \text{Products}$$

$$E \quad + \quad \text{Inhibitor} \qquad \begin{array}{c} E\text{-1 complex} \\ (\text{inactive}) \end{array}$$

Figure 4-6   *Competitive inhibition.*

As we indicated previously, the fact that an enzyme-substrate combination is a specific one has led us to believe that a particular enzyme and its substrate must possess shapes that complement one another if they are to fit together. In other words, the general configuration of the protein molecule must be such that the substrate molecule can fit into a specific site on the enzyme. The specificity and geometry of enzyme reactions are beautifully illustrated by the phenomenon of *competitive inhibition.* To demonstrate this concept, let us consider the enzyme succinic dehydrogenase, which catalyzes the oxidation of succinic acid. If malonic acid, whose structure closely resembles that of succinic acid, is added to the solution, the enzyme's effectiveness is considerably reduced. Experiments show that malonic acid, while not manifesting any change, apparently attaches itself to the enzyme at a position on the molecule that would normally be filled by succinic acid, and therefore competes with succinic acid for the active region of the enzyme molecule.

In Fig. 4-6, the enzyme and substrate are able to achieve an exact "chemical fit," as in the lock and key analogy we have already mentioned. The inhibitor, malonic acid, has a structure similar to that of the substrate and is able to combine at least with part of the active site on the enzyme, thus preventing the substrate from entering. In

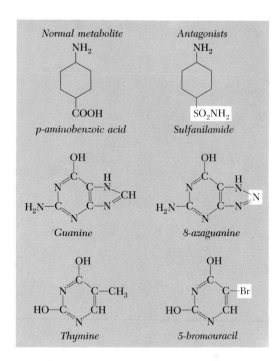

Figure 4-7   *Metabolic antagonists.*

this way, the enzyme is removed from its catalytic role. Since substrate and inhibitor compete for the same site, the degree of inhibition depends on the ratio of substrate to inhibitor. If there is an excess of substrate, the inhibitor may not be able to compete, and no inhibition will be observed. We use this fact to determine whether a substance is acting as a competitive inhibitor; that is, we increase the substrate concentration until the inhibition is removed. Some allowance must be made, however, for the relative affinities of the substrate and the inhibitor for the enzyme. For instance, to obtain a 50 per cent inhibition of an enzyme reaction, we may have to put into the reaction mixture more inhibitor molecules than substrate molecules, for the affinity of the inhibitor for the enzyme may be less than the substrate's affinity for the enzyme.

Competitive inhibitors abound, and some of them are effective antibacterial agents. One of these is sulfanilamide (Fig. 4-7). After the chance discovery that this substance inhibits the growth of certain bacteria, careful analysis revealed that the B vitamin, *p*-aminobenzoic acid (or PABA; see Fig. 4-7), removed the inhibition. Subsequent studies indicated that this was an example of a competitive inhibition.

The success of sulfanilamide and other bacterial killers opened the

Table 4-1 Enzyme inhibitors

INHIBITOR	GROUP ON ENZYME AFFECTED	EXAMPLE OF ENZYME
p-Chloromercuribenzoic acid	—SH	Urease
Iodoacetic acid	—SH and others	Triose phosphate dehydrogenase
Mercury and silver	SH and anions	Glutamic hydrogenase
Cyanide and azide	Metalloporphyrins and metals	Catalase
Carbon monoxide	Metalloporphyrin	Cytochrome oxidase
Citrate, oxalate	Removes metals	Hexokinase
Diisopropylfluorophosphate	Combines with OH group of serine	Chymotrypsin

exciting possibility that all disease-producing organisms, and cancer as well, might be susceptible to the antimetabolite approach, but, unfortunately, the host and disease-producing organisms have very similar enzyme machinery, which means that the altered vitamin or substrate is also quite toxic to the host cells. Since there are some differences, however (in rates, affinities, etc.), between the host and its enemy's enzyme-catalyzed reactions, by careful screening we may be able to find the appropriate chemical that will kill the disease-producing organism before harming the host cells. PABA, for example, is an essential part of folic acid, a key substance for the transfer of methyl ($—CH_3$) groups in both animals and plants. Why certain bacteria are more sensitive to the sulfa drugs than are the host cells is not fully understood at present, but this is a case in which such host-enemy differences exist.

Biochemists commonly employ metabolic antagonists and other enzyme inhibitors to unravel the mechanism of enzyme action. When we speak of a "close fit" between enzyme and substrate, we really mean that there must be specific chemical groups on the enzyme arranged in a particular way. Research efforts, therefore, are directed to finding chemicals that will specifically combine with certain groups on the enzyme. If these groups are important for catalytic activity, inhibition will be observed. In Table 4-1 we list a few of these chemicals and an example of an enzyme which each inhibits. Fortunately, in some cases the inhibitor-enzyme combination is stable, and so we can, by hydrolyzing the protein, determine which amino acid the inhibitor has combined with. Obviously, this type of inhibition is not competitive.

Note that many of the reagents listed in the table are normally classified as poisons. Compounds such as cyanide and carbon monoxide, by combining with the cytochromes (iron-containing catalysts), completely block the electron transport to oxygen. In recent studies, diisopropylphosphofluoridate, or DFP (Fig. 4-8) has been successfully used as an enzyme inhibitor. For example, when the proteinase chymotrypsin combines with DFP, one molecule of inhibitor combines with one molecule of enzyme, causing a complete inhibition of catalytic activity. By degrading the protein-inhibitor complex, we can isolate a very small peptide (6 amino acids) to which the DFP is attached. Such studies seem to indicate that serine and histidine are two amino acids associated with the catalytic site. Interestingly, recent studies have shown that five enzymes with quite different catalytic activity have these same six amino acids associated with the catalytic site.

The current, exciting studies of the mechanism of enzyme action are concentrating on the specific reactive groups on the protein molecule and their relationship to the neighboring amino acids, making it imperative that we know the amino acid sequence in the protein as well as the chemical nature of the active site. In some cases, it has been possible to remove some of the amino acids from the polypeptide without affecting its catalytic activity, providing further evidence that there is an active site and that the entire protein is not essential for catalytic activity.

Catalytic activity is sometimes lost when a small peptide is removed from the protein, but it can be restored to the protein residue by adding equal amounts of the small peptide to the reaction mixture. Since no peptide bonds are formed under these conditions, hydrogen bonding and other noncovalent linkages must be essential in restoring the catalytic site, a fact that reinforces the theory that the secondary and tertiary structures of proteins play a key role in the functioning of enzymes. It will be up to future studies to determine whether any loss in specificity results when part of the polypeptide chain is removed or altered.

Figure 4-8   **Inhibition of an enzyme** *by a specific group reagent.*

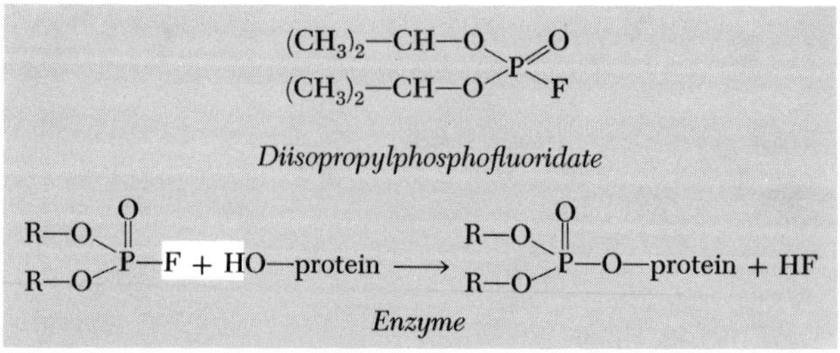

$$(CH_3)_2-CH-O \diagdown \atop (CH_3)_2-CH-O \diagup P \diagup {}^{O} \diagdown_F$$

Diisopropylphosphofluoridate

$$\begin{matrix} R-O \diagdown \\ R-O \diagup \end{matrix} \overset{O}{\underset{\|}{P}}-F + HO-protein \longrightarrow \begin{matrix} R-O \diagdown \\ R-O \diagup \end{matrix} \overset{O}{\underset{\|}{P}}-O-protein + HF$$

Enzyme

We shall be considering many different types of enzymes when we discuss the metabolism of various foods. There are, however, certain points that should be made at this time. In some instances, for an enzyme to function it must be associated with a small molecule called a *coenzyme*. In recent years, biochemists have made the significant discovery that parts of the coenzymes are often specific vitamins, such as riboflavin (B_2) and thiamin (B_1); we shall discuss the function of coenzymes when we consider their need in metabolism.

In addition to the organic cofactors (vitamin-containing compounds), a number of inorganic cofactors are required for enzyme catalyses. For example, iron is necessary for electron transport. Copper, too, apparently functions in electron transport; magnesium is essential for the transfer of phosphate groups, and so on. The mechanisms of action of these various cofactors constitute one of the most active fields in enzyme research today, but much is yet to be learned. It is clear, however, that in some cases the enzyme protein alone is not enough to speed the chemical reactions.

Specific enzymes are extremely useful to the biochemist as tools for determining the structure of biological compounds. But they also have many practical applications. The enzymes best known to the general public are the proteolytic enzymes added to detergents for removing protein stains from textiles. A promising anticancer agent is the enzyme asparaginase isolated from the bacterium *E. coli*.

5 METABOLIC ENERGY

WHEN NUTRIENTS PASS THROUGH THE CELL MEMBRANE, THEY ENTER a new environment: the metabolic machine of the cell. This machine's primary purpose is to convert these nutrients into useful energy and at the same time to create new carbon skeletons that are needed to synthesize other vital compounds. For the past 75 years, research into the cell's metabolism has been a primary activity in biochemistry, and we have now gained a reasonably good understanding of the various reactions in metabolism. Before considering these reactions in detail, however, we must see how energy is liberated in chemical reactions.

For atoms to be joined by specific bonds into molecules, work must be done, i.e., an external source of energy is needed to bring the two atoms together to form a stable molecule. In nature a prime example of energy conversion is sunlight's ability to "put" CO_2 and hydrogen together into high-energy carbohydrates. The energy in the light rays, in effect, is used to make specific chemical bonds. When these bonds are broken, the energy that holds the atoms together is liberated and may be employed to do work. Thus, when an organism, during metabolism, breaks or splits certain atomic links, we want to know whether energy is liberated and whether this energy is released in a form capable of doing useful biological work.

When substance A is converted to substance B, therefore, we must be able to determine whether energy is absorbed or liberated. Suppose we set the energy level of A at some arbitrary value. If energy is given off when A is converted to B, the energy level of B is lower than that of A, and we say it has a minus value. But if energy is absorbed when A is converted to B, then B is said to be energetically more positive than A. If energy is given off, we want to determine whether all or some of it is available to do work on another system. We call this "employable" energy *free energy*. In biochemical systems this free energy is not literally liberated into the environment, but is conserved as special *bond energy* that is passed on to other molecules and used there to form new bonds.

Free energy has two essential characteristics, which can be easily understood by analogy with the energy that can be obtained from water pouring over a dam. The first characteristic is the *distance* the water falls, the second the *quantity* of water that spills over the dam. The product of the two factors equals the total amount of work that can be performed. In his classic studies, Joule let water fall on a paddle from a definite height and measured the temperature change of the water. From this experiment he was able to define the relationship between calories and the mechanical equivalent of heat. A *calorie* is the quantity of heat required to raise the temperature of one gram of water one degree centigrade, starting at 14.5°C (15° calorie) and is equal to 4.185 joules.

Returning to our dam, we see immediately that just a bucket of water will do very little work, even if the distance the water drops is considerable. It takes a sustained flow, and thus a large reservoir above the dam is needed to perform an appreciable amount of work. In other words, a system low in water reserves (or energy potential) has very little capacity to do work.

Let us examine a simple water reservoir as an example (Fig. 5-1). If we open the gate, the water flow from A to B will accomplish work until it is equalized by the flow from B to A. When this reversible process has reached equilibrium, no more work can be gained and the net free-energy change is zero. In a chemical reaction, say when A goes to B, if we know the amount of A that is converted into B up to the time of equilibrium, we can calculate the free energy that has been liberated. Thus a relationship exists between the free energy liberated in a reaction and the extent of the reaction, and we calculate it with the aid of what we call the *equilibrium constant*, which is the ratio of the product, B, and the reactant, A, after the reaction has

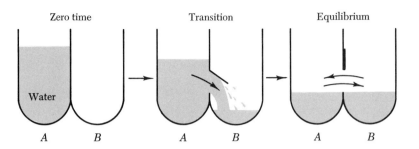

Figure 5-1 **When water flows from one reservoir to another (A → B), work can be performed.** *The amount of energy available depends on the amount of water and how far it falls. At equilibrium the flow from A to B equals the flow from B to A, and no net work is possible.*

reached equilibrium. We have then:

$$\Delta F = -RT \ln K$$

where K is the equilibrium constant (B/A), R is a constant (approximately 2) that is necessary to convert the units into calories, T is the absolute temperature, and ΔF is the change in free energy. If logarithm to the base 10 is used, we multiply the quantities on the right by 2.3 and get:

$$\Delta F = -4.6 \text{ T} \log K$$

If at equilibrium most of A was converted to B, K would be large and ΔF would be large and negative; *the negative sign indicates that energy was liberated.* If very little A is converted to B, the equilibrium constant may be less than 1, and ΔF would be positive, indicating that the reaction favors the formation of A. When energy is liberated in a chemical reaction $(-\Delta F)$, we call it an *exergonic* process. If energy must be supplied to promote the reaction $(+\Delta F)$, we call it an *endergonic reaction.*

The equilibrium constant is an important quantity, for it tells us something about the energy levels of the reactants and products. We may not be able to say precisely how the energy is liberated and possibly transferred to other systems, but we do know from the equilibrium constant what to expect energetically in the bond-breaking process.

In some cases we may not be able to measure the equilibrium constant, and we then have to depend on other methods for determining the energy that is released in a chemical reaction. This is often true for oxidation and reduction reactions. Biologically significant free-energy changes occur in oxidation and reduction reactions, and in these the changes are related to the electrical equivalent of work. In oxidation, electrons are removed; reduction is the reverse process. Since the movement of electrons from one energy level to another (or from one compound to another) is similar to the electrical current in a wire, work can be done by biological oxidations. All we need to know to calculate the free-energy change involved in electron movement is the change in the potential-energy level of the electron.

By studying the capacity of compounds to take up electrons (their oxidizing capacity), we can gain some notion about how they compare as oxidizing or reducing agents. We could, for example, by placing a metal electrode in a solution of a substance, determine whether the substance in question passed electrons to the electrode. If this were the case, we would say that the substance had a higher *electron pressure* (greater reducing power) than the electrode. By taking an electrode and arbitrarily assigning it an electron potential of zero, we can calculate in arbitrary units the oxidizing and reducing strength of a number of compounds of biological interest. From such measurements we can tell how compounds will act relative to one another as oxidants or reductants. When two such systems are coupled together, we can determine the amount of energy liberated or absorbed in the reaction if we know the difference (ΔE) in the potential between the two. This situation is analogous to the water-reservoir system, except that now we use electrons instead of water. The free-energy change produced by the electron flow is calculated from this relationship:

$$\Delta F = n\ 23,000\ \Delta E$$

where n is the number of electrons that move and ΔE is the change in the electrical potential in volts. The constant, 23,000, is necessary to convert the units into calories.

The oxidation of carbohydrates or related compounds is the main source of energy for many organisms. Since organisms are not heat engines, they must liberate the energy in a utilizable form; specifically, they must trap this energy from exergonic processes. Unfortunately, free energy is often considered to be a tangible object, a package of energy that can move from one place to another. Actually, in chemical reactions the energy is first redistributed in the molecule in such a way that when additional molecular bonds are broken, energy is liberated to the environment as heat. Thus organisms must trap the energy-rich intermediate and use it before the energy is lost as heat. An "energy-liberating reaction," therefore, generates an "energy-rich group," which is very reactive and is capable of initiating reactions requiring energy. Energy-yielding reactions must be coupled with energy-consuming reactions in an intimate way if energy is not to be lost as heat. Under these circumstances, the energy is not "liberated" as such but is redistributed in the reacting molecules. In many of these coupled reactions, the over-all process proceeds spontaneously.

The oxidative reactions of the cells are the ones that most actively redistribute or "liberate" energy. Oxidation is any process in which hydrogens or electrons are removed. The simplest type of oxidation is called dehydrogenation. Hydrogen is removed, as occurs in the oxidation of alcohol, $RCH_2OH \rightarrow RCHO + 2H$. Oxidations and reductions cannot proceed, of course, by these half reactions; every substance undergoing oxidation must be accompanied by a substance undergoing reduction, and vice versa. Thus, in the reaction below AH_2 is the reductant and is being oxidized by B.

$$AH_2 + B \rightleftharpoons A + BH_2$$

Since the reaction is reversible, we can start with the chemicals on the right; then BH_2 is the reductant and A is the oxidant. The substances that take part in these reactions have a tendency either to eject or attract electrons, and this tendency is directly proportional to their strength as reducing or oxidizing agents. This is another way of stating what we have already said about the free energy of a reaction: When an electron flows from a reducing system to an oxidizing system, energy is liberated. But exactly how is energy liberated and trapped in an oxidation-reduction reaction?

First let us consider a specific reaction, the oxidation of an aldehyde. We will assume for the moment that we understand how enzymes act, since enzymes are necessary in trapping the energy in the catalytic oxidation of an aldehyde. Enzymes often contain an SH group (sulfhydryl), as is indicated in Fig. 5-2. The aldehyde first combines with the enzyme by splitting the double-bond oxygen in the aldehyde part of the molecule, thus producing a compound that contains an OH group and a single hydrogen on the terminal carbon. One of the double bonds to the oxygen connects with the sulfur in the enzyme molecule, and the hydrogen normally associated with the sulfur moves to the oxygen to satisfy its valence.

In the next step the oxidation process occurs, and energy in the molecule is redistributed (it does not "liberate" energy as is sometimes stated) to form the "energy-rich bond" as indicated.

ENERGY-RICH BONDS

If the link between the carbon and sulfur is broken by water (hydrolysis), a large amount of energy in the form of heat will be liberated. When the molecule is hydrolyzed, it forms an acid (carboxyl group) and the SH enzyme once again. In short, the aldehyde is oxidized in the presence of water to form an acid and a reduced substance. On

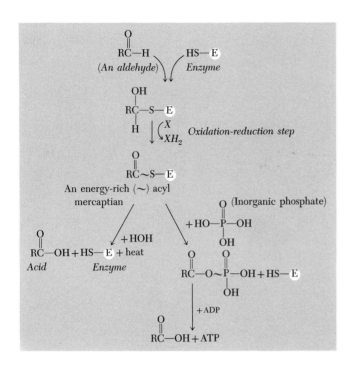

Figure 5-2 *The oxidation of an aldehyde to an acid and the generation of an energy-rich group. Note that in the oxidation of the aldehyde, an energy-rich ($\sim$) intermediate is formed on the enzyme. If water splits this intermediate, the energy is lost as heat, whereas if phosphate splits the intermediate, the energy is conserved in the phosphate bond and can be stored in the form of ATP. (See text for details.)*

the other hand, if the organism is to utilize this energy in the bond it must transfer the energy-rich group to some other molecule, thus synthesizing a new compound by drawing on the energy of the carbon-sulfur link. If the organism does not immediately need the reactive group for synthesis, it can store the energy in some other chemical form, thus freeing the enzyme and conserving the bond energy in a reservoir. *The coupling of this bond energy to other systems is the key reaction in biosynthetic processes.* Instead of hydrolyzing the oxidation product which is called an acyl mercaptan, we can split it with phosphoric acid (H_3PO_4). The phosphate group prevents the loss of energy of the link, resulting in the formation of an energy-rich link between the carbon and phosphorus.

Another important phosphorylated compound, adenosine diphosphate (ADP), whose structure is shown in Fig. 5-3, will react with this energy-rich phosphate group. The product of the reaction is adenosine triphosphate (ATP). ATP has proved to be universally distributed in plants, microorganisms, and animals, and it serves as the initial storehouse for the energy generated in oxidative reactions. This bond energy of the phosphate groups directly or indirectly drives all the energy-requiring processes of life.

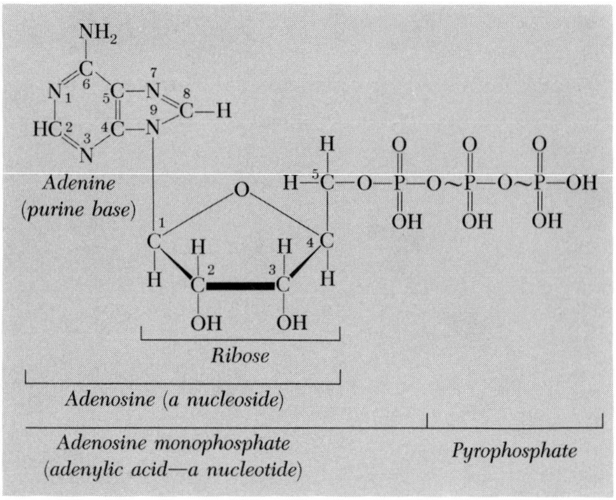

Figure 5-3 **Adenosine triphosphate.** *Note that when the pentose sugar (ribose) is added to adenine, it forms a new compound called a nucleoside. When phosphate is added, it forms the nucleotide. There are a number of such nucleosides and nucleotides that differ in the nature of the base attached to the sugar. The numbering of the various atoms is indicated, i.e., the ribose is attached to N-9 of the adenine and the phosphate is attached to the number 5 carbon atom of the ribose.*

Thus, the energy of an oxidation reaction creates an energy-rich group, and, as we shall see later in our discussion of metabolic processes, this bond energy must couple many reactions that require energy before they can proceed to any significant degree. It turns out that even the second phosphate in ATP is energy-rich, but the third one, which is on the number 5 carbon of the pentose sugar is not energy-rich. ATP, therefore, is capable of supplying two energy-rich units. The link between two energy-rich phosphates, as in ATP, is called a *pyrophosphate bond,* and has proven to be the primary grouping in the conservation of oxidation-reduction energy. We shall discuss numerous examples later when we take up the way the pyrophosphate bond energy of ATP is used in biological reactions. The relationship of ATP energy to biological processes in general is presented schematically in Fig. 5-4.

When ATP is hydrolyzed to liberate the two terminal phosphates, a large amount of heat is liberated—at pH 7.0 the free energy of hydrolysis of the terminal pyrophosphate bond in ATP is approximately 8000 calories, in contrast to the free energy of hydrolysis of an ordinary ester, which may be no more than 1000 or 2000 calories.

Instead of talking about the free energy of a biochemical reaction being *liberated,* therefore, we speak in terms of the creation of bond energy that can be used to drive other reactions. We recognize an energy-rich bond by its ability to couple with other energy-requiring processes. When an aldehyde is oxidized to an acid, the energy does not come off as little packages but is conserved in the form of bond energy. The other way that energy can be redistributed in a molecule so that it can generate an energy-rich phosphate group is by the removal of water; we will study this method when we consider carbohydrate metabolism.

The following simple calculation illustrates the importance of ATP as a source of energy for the cell. The energy requirement for an individual doing moderate exercise is approximately 300 calories per hour. Assuming a 7-hour workday, the total energy requirement is 2100 calories per day. Since 1 mole or 600 grams of ATP supplies 7 calories, 300 moles of ATP are needed to obtain the 2100 calories for a working day. 300 moles of ATP is equal to 180,000 grams or 180 kilograms, which is approximately two to three times the average weight of a man. Assuming 30 per cent efficiency, every individual thus uses and resynthesizes approximately the equivalent of his body weight of ATP each day. Similar approximations can be made for the ATP requirement for other biological processes. For example, a firefly requires the use of 6 micrograms of ATP to produce 100 flashes, while the production of a 1000-volt shock from an electric eel requires 100 milligrams of ATP. For osmotic work, 10 milligrams of ATP are required to concentrate an electrolyte inside the cell thousandfold the concentration outside.

In this discussion of energy metabolism, I do not want to leave the impression that energy is the only significant factor in a given reaction. In the metabolism of a cell, certain key reactions lead to the formation of molecules that are necessary for physiological functions. Although some of these processes may be part of an energy-yielding reaction sequence, the structure of the product may be of considerable importance in initiating other synthetic reactions.

Figure 5-4   All biological reactions depend on the energy derived from converting ATP to ADP.

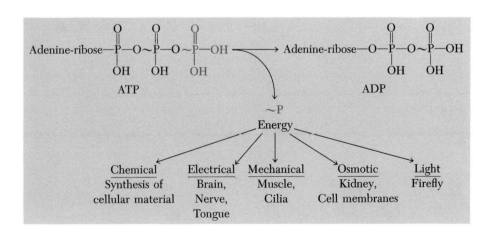

6 ALCOHOLIC FERMENTATION

ALTHOUGH MAN HAS BEEN FAMILIAR WITH ALCOHOLIC FERMENTATION since prehistoric times, he did not discover what causes it until about 1860. A short while before this, however, the French chemist J. L. Gay-Lussac had succeeded in describing the production of alcohol from sugar by this equation:

$$C_6H_{12}O_6 \longrightarrow 2CO_2 + 2CH_3CH_2OH$$

Glucose   Carbon   Ethyl alcohol
                     dioxide

In the middle of the nineteenth century, two schools of controversy developed concerning the mechanism of alcoholic fermentation. One group considered the process to be strictly a chemical one, while the other claimed that it was a process intimately associated with the activities of living organisms. During this controversy, Louis Pasteur was studying how lactic acid is produced in milk and how acetic acid is formed in certain wines. From his experiments he concluded that fermentation takes place in the absence of oxygen (anaerobically) and in the presence of certain microorganisms; when these organisms were excluded no fermentation occurred, thus indicating that fermentation is a physiological process closely bound up with the life of cells—an idea, of course, that contradicted Antoine Lavoisier's earlier conclusion that all living things need oxygen for respiration. In challenging Lavoisier's views, Pasteur suggested that there were substitutes for molecular oxygen and that fermentation, indeed, is the result of life activities being carried on in the absence of air. The experiments and reports of Pasteur did much to establish the belief that fermentation is directly associated with the growth and metabolism of living matter.

More than 20 years elapsed before the next major breakthrough in the investigation of fermentation. In 1897 E. Buchner accidentally stumbled on a discovery that not only opened the door to the secrets of fermentation, but to those of the whole of modern enzyme chemistry as well. Primarily interested in making what are called protoplasmic extracts from yeast to be injected into animals, he ground yeast with sand, mixed it with certain other compounds, and finally squeezed out the juice with a hydraulic press. Since it was difficult to prepare this material each day, he made various attempts to preserve the cell-free extract. Because it was to be injected into animals, ordinary antiseptics could not be used, so he tried the usual kitchen chemistry of adding large amounts of sugar.

When the sucrose was rapidly fermented by the yeast juice, he observed fermentation in the complete absence of living cells for the first time. At last it was possible to study the processes of alcoholic fermentation independently of all other processes such as growth, multiplication, excretion, and the additional complications that accompany fermentation in the living yeast cell. Buchner's work was soon followed by intensive studies of the properties of yeast juice, which was found to be capable of fermenting a large number of sugars such as glucose, fructose, mannose, sucrose, and maltose. Glucose was converted by the juice into ethyl alcohol and carbon dioxide, according to the Gay-Lussac equation.

Before discussing carbohydrate metabolism in detail it is necessary for us to review briefly the chemistry of carbohydrates. Additional details on carbohydrate chemistry may be obtained from the elementary introductory book in this series by White.[*]

Sugars and starches are the principal sources of energy in the ordinary human diet, but they are not essential to the body or to its

[*] E. White, *Chemical Background for the Biological Sciences*, 2nd Ed. (Englewood Cliffs, N. J.: Prentice-Hall, 1970).

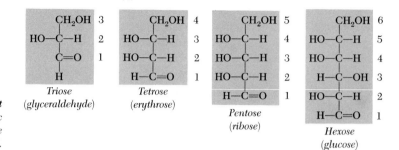

Figure 6-1 Carbohydrates with different numbers of carbon units. Names of specific examples are given in parentheses. The numbering of the carbon atoms is indicated.

individual cells in any way, as far as we know. The cells could obtain their energy just as well from a mixture of protein or fat, or even from other types of carbon-containing molecules. Carbohydrates are the cheapest food commercially, and this is undoubtedly responsible for the large amounts contained in the average diet. Carbohydrates, although used primarily as an energy source, also supply carbon skeletons (carbon atoms linked together and associated with hydrogen and oxygen) that are necessary in the manufacture of the basic components of protoplasm. (For example, glucose, which contains six carbon atoms linked together, can be split to form two compounds containing three carbon atoms each. The three-carbon units can then be used to make other essential compounds that contain only a three-carbon skeleton.)

The basic units in carbohydrates are composed of carbon, hydrogen, and oxygen. A few examples of carbohydrates, of varying chain lengths, are shown in Fig. 6-1. Note that in the specific examples (take glucose, for instance) the OH group can be written on the left or the right side. A three-dimensional model would show this much better, but the main point we wish to emphasize is that many sugars can contain the same number of carbon, hydrogen, and oxygen atoms and differ only in the position of the OH groups. These sugars of the same chemical composition but different chemical properties are called *isomers*. For example, besides glucose the sugars galactose, mannose, and fructose can be represented by the same formula, $C_6H_{12}O_6$.

Since cells are able to discriminate between these isomers, because of the high degree of specificity of reaction shown by the enzyme proteins, the empirical formula of a compound does not give us information of any great value. In order to understand cellular metabolism, we must know the shape of the compounds, and thus we must consider molecular geometry when we discuss the reactions of all classes of compounds. (Note the cyclic form that ribose can take, as indicated below.)

CELL PHYSIOLOGY AND BIOCHEMISTRY

The open-chain form of sugars occurs in aqueous solution but is in equilibrium with the ring form. The ring structure is the usual form that one encounters for longer-chain carbohydrates. The rings are formed from the chain form by the reaction of the hydroxyl group in the 4 or 5 position with the carbonyl group ($-C\underset{H}{\overset{O}{\diagup}}$). The result, as indicated below, is the formation of an oxygen bridge and a ring containing 5 to 6 elements.

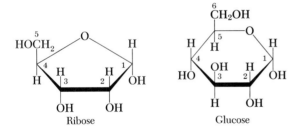

To show the correct steric (spatial) relations of the various groups, the ring structures are drawn as shown below for ribose and glucose. The ring is imagined as being looked at obliquely from above, the three thickened edges being the nearest to the observer.

Ribose   Glucose

Although we will make use of the straight-chain formulae for sugars at various times it should be kept in mind that the ring structure and the stereochemical specificity of enzymes are very important in understanding biochemical processes.

Sugars of the simple type can, like amino acids, be linked together to make longer units. Cane sugar (sucrose), for instance, is made up of a glucose linked to a fructose in a two-sugar chain that is called a *disaccharide*. Strings of three sugars, *trisaccharides*, are common, as are chains of many sugars, or *polysaccharides*. The polysaccharides such as starch and glycogen yield glucose when they are split by water (hydrolyzed).

The first important analysis of the activity of yeast juice was made by Harden and Young in 1905. When they added fresh yeast juice to a solution of glucose at pH 5, fermentation began almost at once. Although the rate of carbon dioxide production soon fell off, they restored it by introducing inorganic phosphate. The recovery was only temporary, however, because the phosphate was soon used up, and the rate of fermentation dropped as the phosphate concentration declined. By adding more phosphate, they sparked another burst of fermentation. Since the inorganic phosphate they added to fermentation mixtures always disappeared, the two scientists suspected that an organic phosphate ester was being formed, and they soon confirmed their suspicions by isolating such an ester in the form of a hexose diphosphate (fructose diphosphate). They observed that this substance was fermented very actively by fresh yeast juice (it was like glucose in this respect) and concluded that it was probably an intermediate in alcoholic fermentation.

Later Robison isolated another sugar phosphate. When examined in detail, the sugar phosphate proved to consist of an equilibrium mixture between glucose-6-phosphate (phosphate attached to the number 6 carbon of glucose) and fructose-6-phosphate. Since both of these sugar phosphates could be fermented, it seemed clear that these substances must arise from a coupling of inorganic phosphate with glucose on carbon number 6. How these esters are formed and in what way the fructose diphosphate is eventually converted into alcohol and carbon dioxide were questions that stumped a number of brilliant biochemists for decades. As we shall see later, studies of muscle extract revealed some of the chemical steps in fermentation, and it gradually became apparent that the fermentation of glucose by yeast juice is very similar to the anaerobic breakdown of glycogen by muscle extracts.

We return to Harden and Young for the next stride forward in our understanding of fermentation. They discovered that yeast juice loses its activity when it is dialyzed, i.e., when the juice is placed in a selective permeable bag surrounded with water. The bag was much like a cellophane membrane, and the tiny holes in it permitted small molecules to move out into the water while the larger molecules stayed inside. Testing the contents of the bag, they found that the dialyzed juice would not ferment but that the activity could be restored by adding the material which had diffused into the water. Additional experiments demonstrated that the contents of the bag were easily destroyed by heating but that the material outside the bag was relatively stable.

From these important experiments, the codiscoverers found that in addition to the large molecules inside the bag, which we now know are enzymes, the yeast juice also contained dialyzable, thermostable substances that are necessary for the function of the enzymes. These dialyzable factors that move away from the catalysts inside the bag are the coenzymes. Thus yeast juice was thought to contain a zymase, which is composed of a nondialyzable thermolabile enzyme, and a cozymase (cofactor), which is dialyzable and thermostable. We now realize, of course, that the zymase is a complex mixture of many enzymes and that several coenzymes are necessary for their function.

At this point we are going to make a big jump in the history of the development of carbohydrate metabolism and consider our present knowledge of the initial steps in glucose metabolism. It is neither necessary nor desirable to give all the details here about the intermediates in the conversion of glucose to alcohol. They are contained in advanced texts and monographs, and some of the prominent ones are listed at the end of the book. We shall only outline some of the key reactions that lead to the formation of alcohol, although we shall occasionally digress to describe the coenzymes or enzymes that are involved.

If glucose and inorganic phosphate are added to dialyzed juice from yeast, no fermentation occurs and no sugar phosphate esters are formed, showing that one of the coenzymes must play a part in the reaction. If ATP is added to the dialyzed juice, however, phosphorylation of the sugar (i.e., the addition of phosphate to the sugar) is begun again, and hexose diphosphate can be isolated. Work with highly purified enzymes has clarified this series of reactions.

The first step in the series consists of the transfer of one of the phosphate groups of ATP to glucose to form glucose-6-phosphate and ADP, in a reaction that is catalyzed by an enzyme, *hexokinase*. (All enzymes concerned with reactions of this type that involve ATP are called *kinases*—fructokinase, glucokinase, etc.) Next, the glucose-6-phosphate molecule, after a series of transformations, reacts with ATP again to form the hexose diphosphate, as is indicated in Fig. 6-2. The hexose diphosphate is now split into two triose phosphate molecules (three-carbon sugars, each of which contains a phosphate). Although the initial products of the splitting reaction are different, they can be converted into one another in a reversible reaction (Fig. 6-3). The crucial intermediate in the formation of alcohol is glyceraldehyde-3-phosphate.

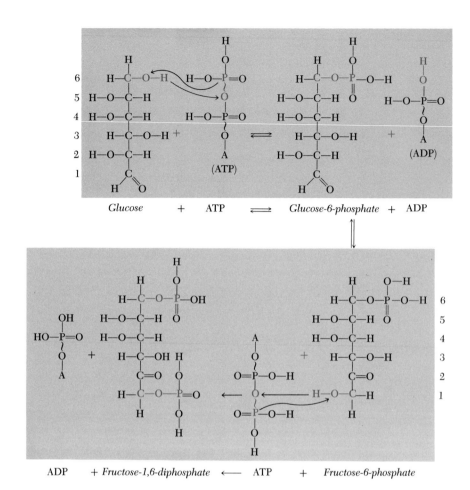

Figure 6-2 **Initial steps in the metabolism** *of glucose-phosphorylation.*

Earlier we saw what happens to an aldehyde when it is oxidized to an acid. In the yeast juice there is an enzyme, triosephosphate dehydrogenase, which catalyzes this oxidative step, in which hydrogen is removed. The glyceraldehyde-3-phosphate adds to the SH group of the enzyme, as shown in Fig. 6-4, and then oxidation occurs to form the energy-rich group on the enzyme. This group, in turn, reacts with inorganic phosphate to form 1,3-diphosphoglyceric acid. In the next stage, the energy-rich phosphate group is transferred from carbon 1 to a molecule of ADP to form 3-phosphoglyceric acid and ATP, a

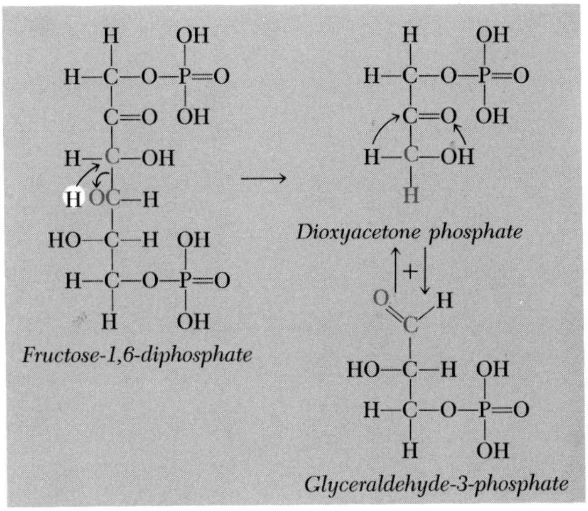

H   OH
H—C—O—P=O
C=O OH
H—C—OH
H OC—H
HO—C—H OH
H—C—O—P=O
H   OH

Fructose-1,6-diphosphate

H   OH
H—C—O—P=O
C=O OH
H—C—OH
H

Dioxyacetone phosphate

O   H
C
HO—C—H OH
H—C—O—P=O
H   OH

Glyceraldehyde-3-phosphate

Figure 6-3   Splitting of fructose-1,6-diphosphate.

reaction that is catalyzed by an enzyme called phosphoglyceric acid kinase, i.e., the reaction is reversible.

The coenzyme of the triosephosphate dehydrogenase plays a vital role as an oxidant in the reaction. As we discussed previously, Harden and Young demonstrated that a small heat-stable cofactor is required for alcoholic fermentation. This dialyzable cozymase (coenzyme 1) was soon isolated and identified as a compound containing adenylic acid, nicotinic acid amide (a B vitamin effective in the prevention of pellagra), and phosphate (Fig. 6-5). These findings provided the first revealing clue to the function of vitamins in cellular metabolism. As we progress into various phases of biochemistry, we shall see what a significant part the vitamins perform in the coenzyme molecules.

After the structure of coenzyme 1 was established, it was referred to as diphosphopyridine nucleotide (DPN).° The reactive group of DPN (and of TPN—triphosphopyridine nucleotide—a coenzyme which differs from DPN by having another phosphate attached to the ribose of the adenylic acid) in biological oxidations is the pyridine ring. This ring can accept two electrons and a proton to form a new structure. In the oxidation of glyceraldehyde-3-phosphate, two hydrogen atoms are removed, but in the reduction of DPN, only one hydrogen ion is added to the ring. Since two electrons are transferred to DPN, a hydrogen remains in the medium (DPN$^|$ + 2H$^+$ + 2e → DPNH

° Recently it has been suggested that DPN be renamed as nicotinamide adenine dinucleotide (NAD).

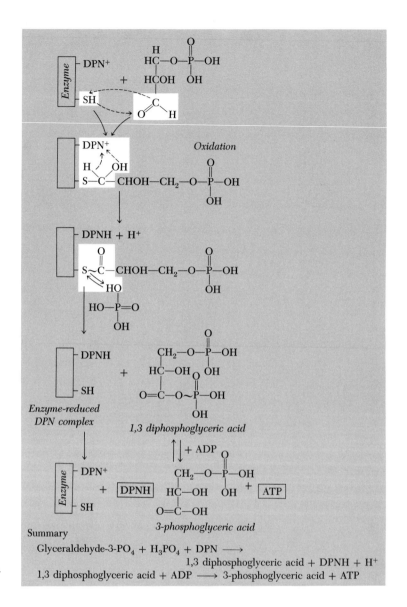

Figure 6-4 **Oxidation of glyceraldehyde-3-phosphate.** *Role of DPN (coenzyme 1).*

+ H⁺), and the reduced DPN is designated as DPNH. In the oxidation of glyceraldehyde-3-phosphate, therefore, the coenzyme (DPN) becomes reduced, 3-phosphoglyceric acid is formed, and ADP is converted into ATP. Once all the available DPN has been reduced, fermentation will cease, but as we shall see later, yeast juice contains other

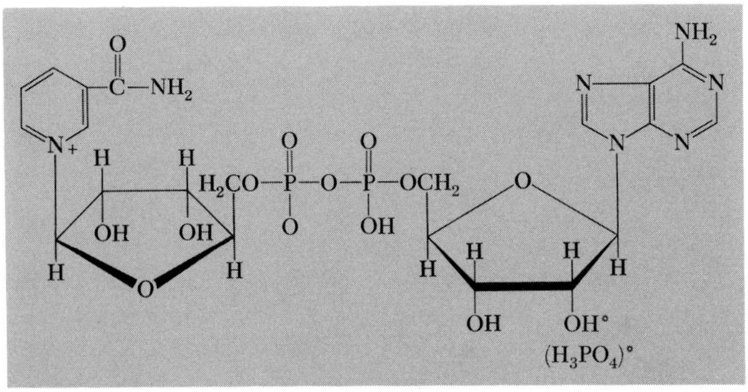

Figure 6-5 **Structure of DPN and TPN.** (*Additional phosphate added to the ribose attached to the adenine.*)

enzymes that catalyze the oxidation of DPNH, thus regenerating the active coenzyme. Note that inorganic phosphate as well as ADP must be continually supplied or regenerated in order to remove the energy-rich intermediate formed on the enzyme surface.

METABOLISM OF 3-PHOSPHOGLYCERIC ACID

The yeast juice further metabolizes the 3-phosphoglyceric acid in the following manner. The 3-phosphate ester is converted into the 2-phosphate ester (2-phosphoglyceric acid), which is then dehydrated, causing the energy in the molecule to be redistributed into an energy-rich phosphate group, phosphoenolpyruvic acid (Fig. 6-6). The formation of phosphoenolpyruvic acid from 2-phosphoglyceric acid is catalyzed by an enzyme called *enolase.* In the presence of phosphate, enolase

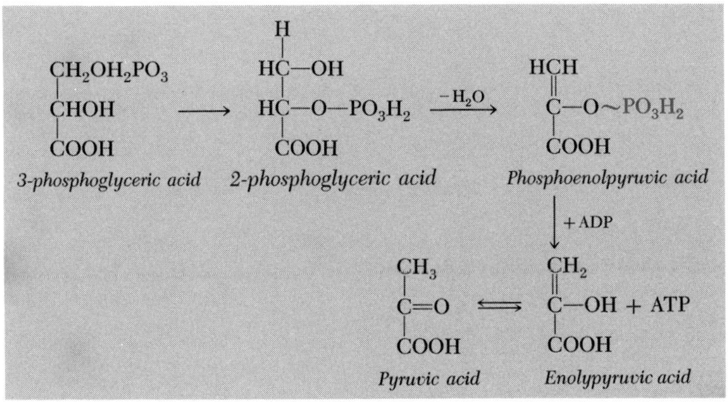

Figure 6-6 ***Metabolism of 3-phosphoglyceric acid.***

is strongly inhibited by fluoride, because of the removal of magnesium (as magnesium fluorophosphate), which is essential for the action of enolase. If fluoride is used in a yeast juice fermentation, then, the metabolism of 2-phosphoglyceric acid can be prevented, a fact that enabled biochemists to accumulate 3-phosphoglyceric acid in quantities that could be isolated and identified. They were then able to demonstrate that when 3-phosphoglyceric acid is added to an uninhibited yeast extract, it is rapidly converted into phosphoenolpyruvic acid.

This new energy-rich phosphate group is no different from the one formed in the oxidation of glyceraldehyde-3-phosphate. The phosphate in the presence of a specific kinase can be transferred to ADP to form ATP and pyruvic acid. Pyruvic acid is an example of an α-keto acid. The first step in the metabolism of such acids is the loss of carbon dioxide (decarboxylation). In yeast extracts, the decarboxylation reaction leads to the production of carbon dioxide and acetaldehyde ($CH_3COCOOH \rightarrow CH_3CHO + CO_2$). The enzyme that catalyzes this reaction is called carboxylase and requires as a cofactor (cocarboxylase) another one of the B vitamins, thiamine. For thiamine to be enzymatically functional, it must be phosphorylated to form thiamine pyrophosphate (TPP). In the final stage of the production of alcohol, acetaldehyde is reduced to alcohol by the hydrogens that were passed on to DPN in the oxidation of glyceraldehyde-3-phosphate. The reduction of the acetaldehyde to alcohol or the oxidation of alcohol to form acetaldehyde is catalyzed by an enzyme called alcohol dehydrogenase. The coenzyme necessary for the action of this enzyme is DPN ($CH_3CHO + DPNH + H^+ \rightleftarrows CH_3CH_2OH + DPN$).

A summary of these remarkable reactions that are catalyzed by yeast is shown in Fig. 6-7. The unraveling of each step, the isolation and purification of the numerous enzymes and coenzymes, and the establishment of the scheme for the whole yeast cell represent some of the most significant and magnificent developments in the history of biochemistry and cellular physiology. The over-all results of this series of reactions, starting with glucose, are:

First, for each molecule of glucose fermented two molecules of alcohol and two of carbon dioxide are formed.

Secondly, for each molecule of glyceraldehyde-3-phosphate oxidized, one molecule of DPN is reduced and later reoxidized at the expense of the molecule of acetaldehyde that is formed from the decarboxylation of pyruvic acid. Two triose molecules are produced from each glucose molecule, leading to the formation of two molecules of alcohol and CO_2.

Thirdly, two molecules of ATP are employed in the phosphorylation of each molecule of glucose. The second ATP is used in the

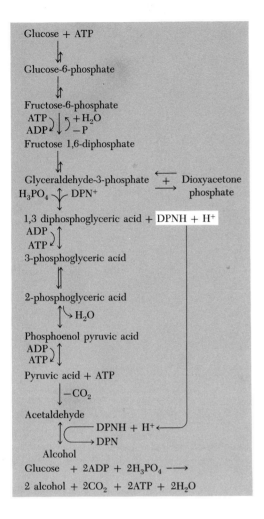

Glucose + ATP

Glucose-6-phosphate

Fructose-6-phosphate
ATP $+H_2O$
ADP $-P$

Fructose 1,6-diphosphate

Glyceraldehyde-3-phosphate $\xleftarrow{+}$ Dioxyacetone
H_3PO_4 DPN^+ phosphate

1,3 diphosphoglyceric acid + $\boxed{DPNH + H^+}$
ADP
ATP

3-phosphoglyceric acid

2-phosphoglyceric acid
H_2O

Phosphoenol pyruvic acid
ADP
ATP

Pyruvic acid + ATP
$-CO_2$

Acetaldehyde
$DPNH + H^+ \leftarrow$
DPN

Alcohol

Glucose + 2ADP + $2H_3PO_4 \longrightarrow$
2 alcohol + $2CO_2$ + 2ATP + $2H_2O$

Figure 6-7 *Summary of the reactions in alcoholic fermentation.*

formation of hexose diphosphate, each molecule of which yields two molecules of glyceraldehyde-3-phosphate; each of these takes up a molecule of inorganic phosphate after it has been oxidized to form two energy-rich phosphate groups that are transferred to ADP to produce ATP. Therefore, up to this stage in fermentation the yeast has recovered only the amount of ATP used in the initial stages. In the dehydration of the two molecules of 2-phosphoglyceric acid, however, two additional energy-rich phosphate groups are formed. In alcoholic fermentation there is thus a net synthesis of two molecules of ATP for each molecule of glucose metabolized.

ALCOHOLIC FERMENTATION

Note that the splitting of pyruvic acid is probably the only irreversible reaction in the entire fermentation process. The conversion of fructose-1,-6-diphosphate back to fructose-6-phosphate does not lead to the reformation of ATP. The "reversibility" of this reaction depends on the removal of the phosphate in the number 1 position by hydrolysis. The reaction is catalyzed by a specific enzyme (phosphatase). Note the importance of inorganic phosphate and ADP as controlling factors in cellular metabolism.

FOR A LONG TIME AFTER THE DISCOVERY OF THE PHOSPHORYLATION OF hexoses in alcoholic fermentation by Harden and Young, the process was not considered significant except as a means of shaping the hexose molecule for fermentative breakdown. Similar studies of other cells and particularly of muscle, however, revealed that phosphate plays a dominant role in energy transformation, especially in that involved in muscle contraction. The questions confronting the early investigators of muscle contraction were: What is the chemical source of the energy that powers the muscle machine, and how is this energy transformed into the mechanical energy of contraction?

The investigators soon uncovered these facts: (1) muscle can contract in a normal manner in the complete absence of oxygen; (2) lactic acid is produced during anaerobic contractions and accumulates with continued stimulation until the muscle becomes fatigued; (3) if the fatigued muscle is then put into oxygen, it recovers its ability to contract, and lactic acid simultaneously disappears; (4) less lactic acid is formed in a muscle that has access to oxygen than in one that contracts anaerobically. All the information pointed to the existence of a pro-

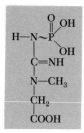

Figure 7-1   Creatine phosphate.

portionality between the amounts of work done, of heat produced, of tension developed in a muscle, and the quantity of lactic acid formed.

By the late 1920's, it became evident that some of the energy expended in anaerobic muscle contraction comes from the conversion of the polysaccharide glycogen (which is stored in muscle) to lactic acid in a process called *glycolysis*. For many years, physiologists and biochemists thought that the energy-supplying glycogen breakdown occurred just when the muscle contracted, but A. V. Hill found, after carefully measuring the heat during the contraction of a single muscle fiber, that there were two kinds of heat in addition to the initial heat associated with contraction: a relaxation heat and a delayed heat output called recovery heat. When he placed muscle under anaerobic conditions, however, he observed the contraction and relaxation heat but no recovery heat, and concluded that oxygen was needed to obtain the recovery heat. Since lactic acid accumulated in his experiment, there still appeared to be a close correlation between glycolysis and muscle contraction.

At about this time, a new chemical substance was isolated from muscle, a substance that in acid solutions readily broke down into creatine and phosphate. The compound turned out to be creatine phosphate and to be energy-rich just like the phosphate in ATP (see Fig. 7-1). Since it seemed to disappear during muscle contraction, it was thought to be related in some way to the energy supply and glycolysis. When the pure compound became available, large amounts of heat were found to be liberated when the compound was hydrolyzed, thus proving conclusively that a great quantity of potential energy was stored in the nitrogen-phosphorus bond. But one still had to admit that muscle contraction seemed to be closely connected with the breakdown of glycogen into lactic acid.

In 1930 the puzzle began to clear up when A. Lundsgaard demonstrated that muscle contraction could proceed anaerobically without the formation of lactic acid, provided he poisoned the muscle with a compound called iodoacetic acid. He reported that this nonlactic acid contraction was accompanied by a pronounced breakdown of creatine phosphate, and when the latter was exhausted the muscle would no longer contract. From earlier work, the amount of heat produced when creatine phosphate is hydrolyzed was known. When researchers measured the quantity of creatine phosphate broken down during muscle contraction and compared it with the expected amount of heat, they found that an exact relationship existed between the amount of creatine phosphate broken down and the amount of heat produced. In other words, when muscle contracts anaerobically, without glycogen breakdown, the heat production can be accounted for by the breakdown of creatine phosphate.

Investigation at this time into the biochemistry of the breakdown of glucose and glycogen by mammalian tissues, particularly by muscle cells, revealed that intermediates in the breakdown of these substances are essentially the same as those found in alcoholic fermentation. Lundsgaard was able to trigger muscle contractions without producing lactic acid, because the iodoacetic acid combined with the SH group in the triose phosphate dehydrogenase and prevented the oxidation of glyceraldehyde-3-phosphate. Creatine phosphate thus disappeared as a result of the inhibition of ATP synthesis. If dialyzed muscle extract were used, creatine phosphate was broken down only when ADP was present, indicating that creatine phosphate did not directly react with the muscle proteins. Subsequent work by Englehardt showed that the muscle protein (ATPase) reacted with ATP to break it down into ADP and inorganic phosphate. This was the first clear suggestion that ATP might be the immediate energy source for muscle contraction. Additional evidence demonstrated that creatine phosphate (CP) acts as a reservoir of high-energy phosphate which can rephosphorylate ADP to form ATP in this reversible reaction:

$$CP + ADP \rightleftharpoons ATP + C$$

The reaction is catalyzed by an enzyme called creatine phosphokinase. ATP is thus generated in the glycolytic process just as it is in alcoholic fermentation, and when excess ATP is available, the high-energy phosphate is transferred to creatine. When a muscle that has been poisoned by iodoacetic acid contracts, the ATP can be generated only at the expense of this creatine phosphate reservoir.

Although much is known about the contractile proteins of muscle, the mechanism of converting the chemical energy of ATP into mechanical work remains obscure. In the chemical composition of muscle, there are two important contractile proteins, actin and myosin. When actin and myosin are mixed they form a complex (actomyosin), which can be made into "synthetic" threads that rapidly decompose ATP and at the same time are capable of contracting. We can be reasonably certain, then, that actin and myosin are the major proteins of the contractile muscle fiber. The mechanical arrangements of these fibers in the muscle and the function of ATP in causing contraction are gradually being clarified, particularly for skeletal muscle. The present view is that the contraction is produced by the sliding of actin and myosin fibers over one another. Since ATP under appropriate conditions is known to dissociate or to separate actin and myosin, when the nerve impulse stimulates a muscle to contract, the first event may be the activation of the ATPase system. Destruction of the ATP would allow re-association of the actin and myosin, resulting in a contraction,

Figure 7-2 Schematic representation showing the relationship between actin and myocin filaments from muscle and what is thought to occur when ATP energy induces contraction.

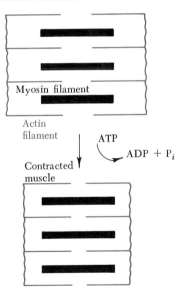

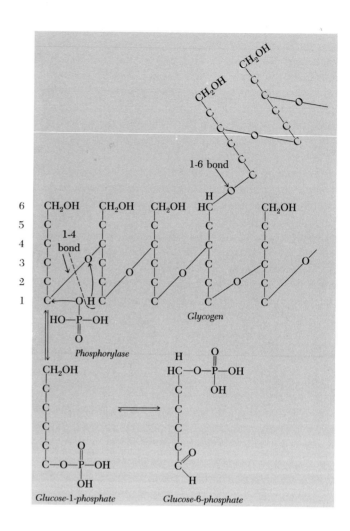

1-6 bond

1-4 bond

Phosphorylase

Glycogen

Glucose-1-phosphate

Glucose-6-phosphate

Figure 7-3   Initial step in glycogen metabolism.

without changing the shape of the individual fibers. The schematic relationship between the actin and myosin is shown in Fig. 7-2.

Glycogen is a complex polymer of glucose. As shown in Fig. 7-3, the sugar molecules are linked together by a bond (glycosidic) between the number 1 carbon of one glucose and the number 4 carbon of the next glucose. Branching from this straight chain, which may have as many as 18 glucose units, are a number of 1–6 glycosidic linkages that then continue as 1–4 linkages. Glycogen, therefore, is a branched, complex polysaccharide with a molecular weight as high as four million. Enzymes attack it at either the 1–4 or the 1–6 linkages. Many hydrolytic enzymes (glycosidases-amylases) can break down glycogen to smaller polysaccharides or free-glucose units. In muscle, when glycogen is broken down it reacts first with inorganic phosphate instead of water at the 1–4 glycosidic linkage; the process is thus a phosphorolysis, not a hydrolysis.

The enzyme that catalyzes this reaction is phosphorylase, and the product of the reaction is glucose-1-phosphate and the remainder of the glycogen molecule. Once the outer tier of 1–4 glycosidic linkages is broken, the reaction stops. There are hydrolytic enzymes, however, that split the 1–6 linkage and thereby expose additional 1–4 glycosidic linkages which phosphorylase will attack. The new phosphate ester formed in the breakdown of glycogen is readily converted into glucose-6-phosphate by the enzyme phosphoglucomutase; this phosphate ester, you will recall, is the compound formed when ATP reacts with glucose. All these reactions are reversible, and if there is an excess of glucose and ATP, glycogen synthesis will occur.

The control of synthesis and degradation of glycogen is an intriguing and informative chapter in the history of biochemistry. Muscle phosphorylase exists in two forms, a and b. Phosphorylase a has a molecular weight of 500,000 and is made of four identical polypeptide chains. Each polypeptide contains a serine residue whose hydroxyl group is esterified to phosphate. Muscle also contains a specific phosphatase that will remove these phosphates. When this happens phosphorylase a dissociates into two inactive dimeric phosphorylase b molecules: Phosphorylase b can be converted into phosphorylase a by using ATP to phosphorylate the serine residues as shown in the following equation:

$$2 \text{ phosphorylase b} + 4\text{ATP} \longrightarrow \text{phosphorylase a} + \text{ADP}$$

The interconversion of phosphorylase a and b (presented schematically in Fig. 7-4) is the basic mechanism for the control of glycogen breakdown. The conversion of phosphorylase b into a requires the presence of an active phosphorylase b kinase. The formation of this active kinase depends upon a second kinase that requires a specific cofactor, 3′,5′ cyclic adenylic acid (see Fig. 7-5). Cyclic AMP is formed from ATP by the enzyme adenyl cyclase whose activity is under hormonal control by epinephrine. Thus, the sequence of events upon the release of epinephrine from the adrenal glands, which occurs when one is frightened, leads eventually to the formation of active phos-

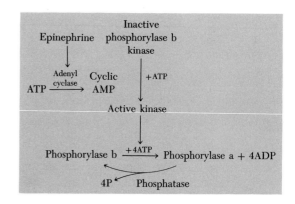

Figure 7-4   *Interconversion of phosphorylase a and b:* Control of glycogen breakdown.

phorylase a. This in turn stimulates the breakdown of glycogen leading to the rapid synthesis of ATP.

Cyclic AMP appears to be important in regulating a number of cellular processes in various stages of growth and development. This often occurs by way of hormones which appear to effect the concentration of cyclic AMP in the target organ. Thus the hormone is the first messenger which travels to the target cells where it increases the formation of the second messenger, cyclic AMP. Activities which are affected by cyclic AMP include in addition to phosphorylase, phosphofructokinase, cell aggregation in slime mold, HCl secretion in the gastric mucosa, release of insulin, kidney permeability and others.

The synthesis of glycogen follows a pathway different from the reversal of the phosphorylase reaction; however, glucose-1-phosphate is the initial substrate. The second substrate is a compound very similar to ATP, namely uridine triphosphate (UTP). As shown in Fig. 7-6, UTP reacts with glucose-1-phosphate to form uridine diphosphate glucose (UDPG) and inorganic pyrophosphate. UDPG is referred to as "active" glucose, since in the presence of the enzyme glycogen synthetase the glucose of UDPG is transferred to glycogen to increase the chain length by way of 1,4 glycosidic linkages. Glycogen synthetase occurs in a phosphorylated and dephosphorylated form. The dephosphorylated enzyme is enzymatically active, and can be converted into the inactive form by a specific kinase and ATP. This inactive enzyme can be activated by glucose-6-phosphate, thus assuring glycogen synthesis only if excess sugar and phosphate bond energy are available (UTP). The glycogen synthetase kinase, like the phosphorylase b kinase is active only in the presence of cyclic AMP; thus, epinephrine also affects glycogen synthesis. The relationship between glycogen synthesis and breakdown is shown in Fig. 7-7.

Figure 7-5   *Structure of cyclic adenylic acid.*

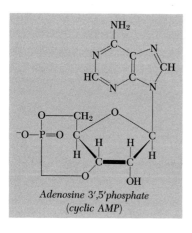

Adenosine 3',5'phosphate
(cyclic AMP)

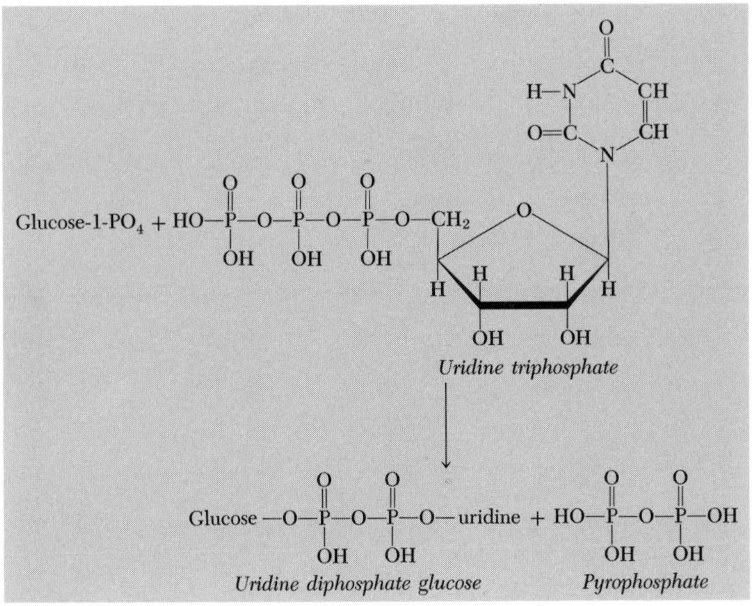

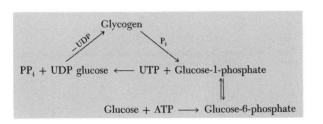

Figure 7-6 *Formation of uridine diphosphate glucose from UTP and glucose-1-phosphate.*

After the formation of glucose-6-phosphate, glycolysis and alcoholic fermentation follow a common path until pyruvic acid is formed. Here the pathways again diverge, for muscle, unlike yeast, does not contain carboxylase, the enzyme that decarboxylates pyruvic acid. Pyruvic acid will, nevertheless, react with the reduced DPN and oxidize it in much the same way as the acetaldehyde did in the yeast extract. When pyruvic acid is reduced by DPNH, lactic acid is formed, and the enzyme that catalyzes this reversible oxidation-reduction reaction is called lactic dehydrogenase. The net effect of the anaerobic reaction sequence in muscle is that for every glucose unit (from glycogen), two molecules of lactic acid are produced. DPN is alternately reduced and oxidized, as it is in alcoholic fermentation.

Figure 7-7 *Synthesis and breakdown of glycogen.*

In yeast juice, as in muscle extract, the sequence generates four new energy-rich phosphate bonds for each glucose molecule metabolized. In the fermentation of free glucose, however, two of the new bonds are used in the preliminary phosphorylation of the glucose molecule, so there is a net gain of only two molecules of ATP. In the first stage in glycolysis, on the other hand, the glycogen is split by inorganic phosphate and not by ATP, so that when the product, glucose-1-phosphate, is converted into glucose-6-phosphate, only one molecule of ATP is needed to produce each molecule of hexose diphosphate. Muscle glycolysis, therefore, gains a net of three molecules of ATP for each glucose unit of glycogen metabolized, enabling us to classify glycogen as an energy reservoir. Although the ester link of glucose phosphate that is formed in the uptake of free glucose uses

Figure 7-8 **Some products of fermentative metabolism.**

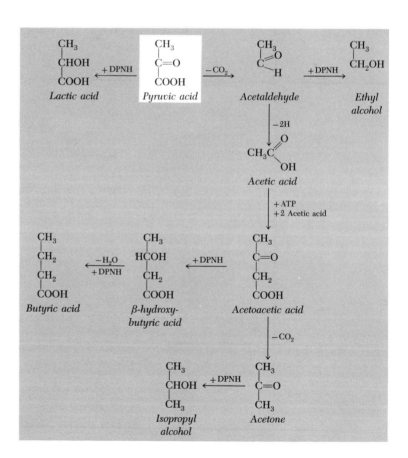

ATP, the energy is not lost when the glycosidic linkage in glycogen is created, for this energy is such that inorganic phosphate can react with it to form again a hexosephosphate ester, which can be metabolized.

The control of glycogen or glucose metabolism in muscle is very similar to the process described for yeast. The reoxidation of DPNH, the action of inorganic phosphate, and the utilization of ATP are the essential processes. Can you describe now the chemical events that take place in muscle during a 440-yard dash? What happens to ATP? If the inorganic phosphate increases in concentration, what happens to glycogen breakdown?

In muscle glycolysis, then, the same fundamental mechanisms—oxidation-reduction, dehydration, and phosphorylation—are in operation as are present in alcoholic fermentation. These reactions are the basic ones involved in energy and carbon transformation in fermentations. In the past 35 years, we have found that in the metabolism of various organisms, these relatively simple steps account for almost all the fermentation products formed from carbohydrates by the organisms. For example, lactic acid bacteria produce lactic acid by a sequence of reactions identical to those observed in muscle. Outlined in Fig. 7-8 are a number of reactions that are carried out by different organisms. Note that in the formation of most of these compounds, only simple oxidation-reduction, hydrolytic, or decarboxylation reactions occur, and their net effect is to produce a number of electron acceptors that then oxidize the reduced DPN, which is formed in the triose phosphate dehydrogenase reaction. Pyruvic acid is formed in the same way by these organisms.

Thus far we have discussed the oxidation reactions in which DPN acts as the electron acceptor. This step is the important initial dehydrogenation in the metabolism of a number of substrates. In a series of brilliant studies beginning in the early 1920's, Otto Warburg found traces of two other important coenzymes in cells grown under aerobic conditions. He was intrigued by the ability of iron compounds—blood charcoal in particular—to *catalyze* the oxidation of many different organic substances, using molecular oxygen as the electron acceptor. Pure charcoal made by heating sucrose does not possess this property, and Warburg attributed the catalytic action of blood charcoal to its iron content. From this and other evidence, he concluded that an iron-containing substance in the cell is essential for the activation and utilization of oxygen.

THE OXIDATION OF REDUCED DPN
BY OTHER OXIDANTS

The importance of iron compounds in oxygen utilization was also suggested by the observation that low concentrations of cyanide and carbon monoxide inhibit respiration. Warburg found that cyanide blocked iron-catalyzed oxidations and concluded that it affected cellular respiration by combining with an iron-containing "respiratory enzyme" (Atmungsferment). In addition, he observed that the respiratory inhibition by carbon monoxide could be reversed by visible light. Since earlier experiments by others had shown that carbon monoxide would combine with hemoglobin but could be dissociated with visible light, Warburg concluded that his respiratory enzyme was an iron compound very much like hemoglobin.

Subsequent studies by Keilin clearly revealed the importance of a number of hemes, or iron-containing pigments, for electron transport. He called these cellular pigments *cytochromes* and was able to demonstrate at least three different heme proteins, which he named cytochromes a, b, and c respectively.

Warburg and Christian also isolated from cells a second coenzyme that was needed to set off oxidative reactions. They were able to isolate from yeast a yellow protein capable of catalyzing the transfer of electrons from reduced DPN to some other acceptor. They removed the yellow component by dialysis and prepared it in pure form. It proved to be a riboflavin compound that could accept electrons from DPNH. Since reduced flavins are colorless, the two investigators followed the reduction process by observing the disappearance of the yellow color. As might be expected, it is not the free riboflavin that is effective in electron transport, but the nucleotide derivative.

Two forms of riboflavin involved in electron transport occur in nature. One is flavin mononucleotide or riboflavin phosphate, and the other is flavin adenine dinucleotide (Fig. 7-9), which is analogous to DPN, a substance formed from another B vitamin, nicotinic acid amide. The flavins can exist in either the oxidized or the reduced form, and DPN is one of the better reducing agents in the presence of certain enzymes. Although reduced flavins are readily reoxidized by oxygen when they are free in solution, they are often poorly oxidized when bound to an enzyme. The reasons for this are not entirely clear.

The flavin operates as an intermediate carrier of hydrogen between reduced DPN and other electron acceptors before oxygen is reduced. A number of outstanding investigations have indicated that the *cytochromes* are the oxidizing agents for the reduced flavins. Cytochrome pigments are present in all aerobic cells and can be alternately reduced and oxidized by a number of agents. Extensive studies have proved that the cytochromes contain a protein and an iron porphyrin group (Fig. 7-10). The iron porphyrin of the cytochromes is made up of four

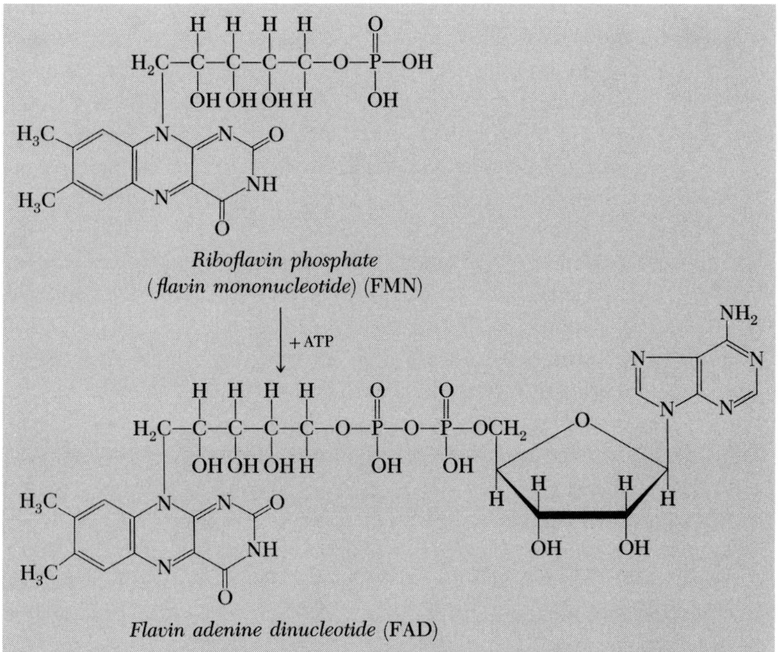
Riboflavin phosphate
(flavin mononucleotide) (FMN)
+ATP
Flavin adenine dinucleotide (FAD)

$$DPNH + FMN \longrightarrow DPN^{+} + FMNH_2$$

$$\text{or} \qquad\qquad\qquad \text{or}$$

$$FAD \qquad\qquad\qquad FADH_2$$

Figure 7-9 ***Riboflavin coenzymes.***

Figure 7-10 ***Cytochrome C.***

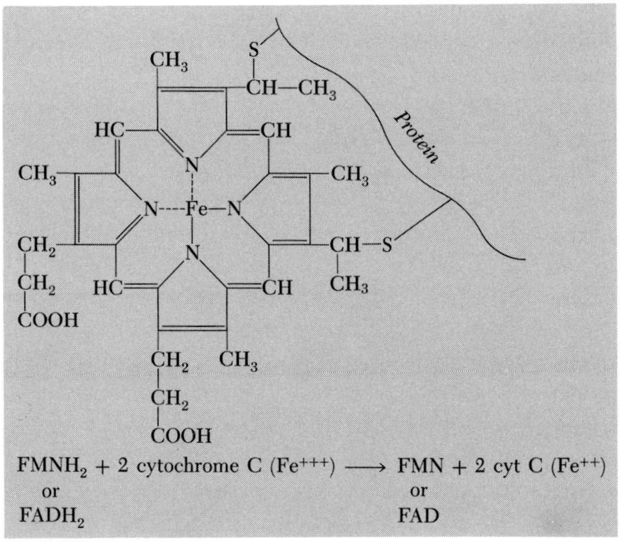
Protein

$$FMNH_2 + 2\ \text{cytochrome C}\ (Fe^{+++}) \longrightarrow FMN + 2\ \text{cyt C}\ (Fe^{++})$$

$$\text{or} \qquad\qquad\qquad \text{or}$$

$$FADH_2 \qquad\qquad\qquad FAD$$

pyrrole rings and an atom of iron, as shown in Fig. 6-5, and is very similar to the pigment in hemoglobin. Cyanide and carbon monoxide are potent poisons of these iron-containing catalysts. In cytochromes, it is the iron in the porphyrin that undergoes oxidation-reduction: $Fe^{+++} + e \rightarrow Fe^{++}$. One of the cytochromes, cytochrome C, appears to be particularly important in the transfer of electrons to oxygens. The porphyrin of this cytochrome attaches itself to a protein by forming a linkage with two sulfur atoms in the cysteine residues of the polypeptide chain. An additional enzyme is necessary to oxidize the reduced cytochrome C by molecular oxygen and is called *cytochrome C oxidase*. The evidence indicates, therefore, that the mechanism of the electron transport to oxygen is from the flavin coenzymes to the cytochromes and then to molecular oxygen. In all aerobic organisms studied, the pathway of electron flow seemed to occur in the following sequence: DPN → flavins → cytochromes → molecular oxygen.

CARBON DIOXIDE AND THE FORMATION OF ELECTRON ACCEPTORS

Succinic acid $(COOHCH_2CH_2COOH)$, a four-carbon dicarboxylic acid, is manufactured in a number of organisms. For many years its formation puzzled biochemists, since it was not obvious how a substance containing four carbon atoms could arise from a six-carbon sugar or a three-carbon compound such as pyruvic acid. Earlier work indicated that the CO_2 tension in a medium containing growing cells tends to increase the productivity of succinic acid. While studying glycerol fermentation in propionic acid bacteria, Wood and Werkman first established that CO_2 uptake (fixation) in nonphotosynthetic organisms is an essential process in the formation of key compounds that can accept electrons from DPNH.

Subsequent isotope work with C^{13}- and C^{14}-labeled CO_2 revealed that when carbon dioxide is in some way added to the methyl group of pyruvic acid, a four-carbon acid called oxalacetic acid is formed. The oxalacetic acid, when reduced with DPNH, forms malic acid. When it loses water, malic acid gives rise to fumaric acid which can be reduced by DPNH to produce succinic acid. The sequence of reactions is shown in Fig. 7-11. Thus CO_2, by combining with pyruvic acid, forms key compounds that can act as electron acceptors.

When Szent-Györgyi in his early experiments found that compounds such as succinic and fumaric acid will stimulate respiration of tissues, he considered them to be significant in the hydrogen transport process and thus assigned them a catalytic role. The function of these four carbon dicarboxylic acids in electron transport remained a mystery, however, until Krebs discovered that they are involved in a cyclic

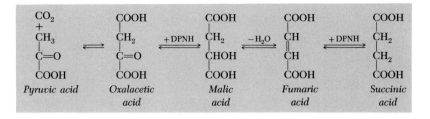

Figure 7-11   Formation of succinic acid.

process of electron transport. The catalytic effect of these acids on respiration, explained Krebs, results from their participation in a cyclic series of reactions that is needed for the oxidation of pyruvic acid to CO_2 and water. In these reactions the electrons are transported to oxygen over the DPN-flavin-cytochrome system. The net effect of this series of reactions is the complete combustion of pyruvic acid to CO_2 and water.

8 OXIDATIVE METABOLISM

CONSIDERING THE NUMBER OF GLUCOSE UNITS USED, GLYCOLYSIS IS AN inefficient mechanism for the synthesis of ATP, since a great deal of energy is still available in the fermentation products. As we indicated in the last chapter, the oxidative metabolism of pyruvic acid is the most effective method of generating phosphate-bond energy in aerobic organisms. This stage of metabolism of carbohydrates is called the Krebs citric acid cycle, and it stands at the intersection where the main routes of cellular metabolism converge. This cyclic metabolic machine handles products derived from the metabolism not only of glucose but also of amino acids and fatty acids.

PYRUVIC ACID OXIDATION

In one of the initial key reactions, the metabolism of pyruvic acid, carbon dioxide is removed and an "active" (energy-rich) two-carbon unit is formed. This initial reaction of pyruvic acid is similar to the one that occurs in the formation of alcohol, since the first step is a decarboxylation of pyruvate to form an active aldehyde. The main

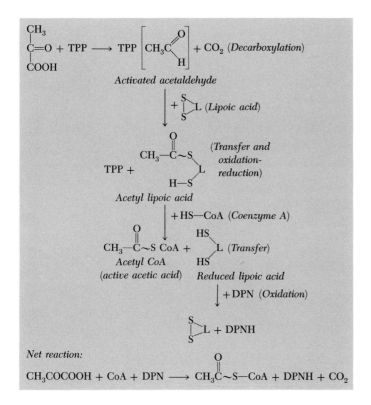

Figure 8-1   *Oxidative decarboxylation of pyruvic acid.*

difference here, however, is that the two-carbon fragment formed is not free acetaldehyde, but is combined to the coenzyme (thiamine pyrophosphate—TPP) of the decarboxylase (Fig. 8-1). This activated aldehyde is transferred to a second coenzyme, lipoic acid, which contains a disulfide (S—S) link. In this transfer, the S—S bond is reduced by the active aldehyde, leading to the formation of an SH group and the energy-rich acetyl group attached to the second sulfur atom.

The active two-carbon unit is similar to the three-carbon energy-rich unit that forms in the triose phosphate dehydrogenase reaction. This active C_2 unit is now transferred to an SH group in another coenzyme (coenzyme A, or CoA), resulting in the formation of completely reduced lipoic acid (2SH) and acetyl coenzyme A. The reduced lipoic acid reacts with DPN to form reduced DPN and the original active form (S—S) of the lipoic acid molecule. The net result of this series of rather complicated reactions (outlined in Fig. 8-1) is the formation of an energy-rich two-carbon unit, DPNH and CO_2. Note that the energy liberated in this *oxidative decarboxylation* is not lost

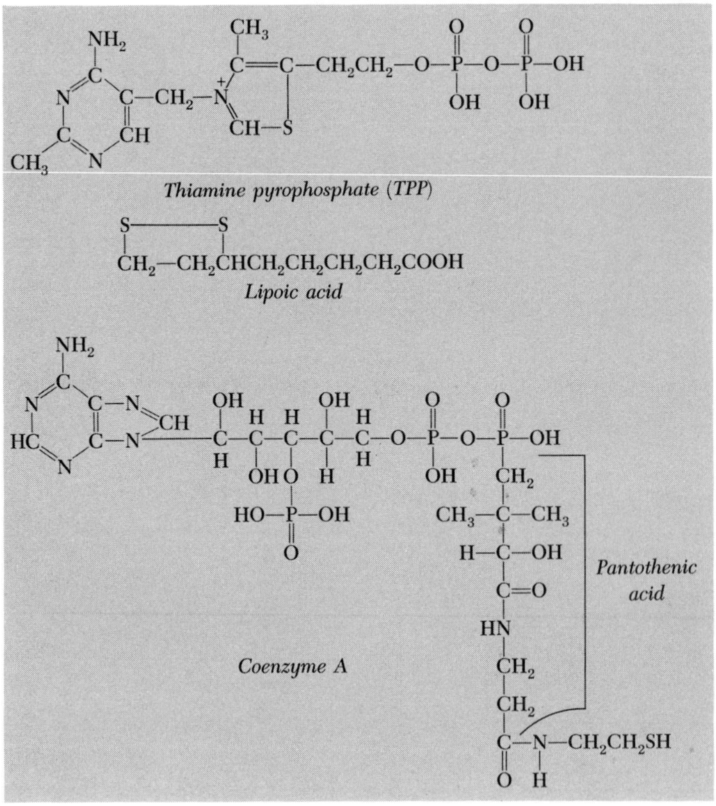

Figure 8-2 **Coenzymes** *for pyruvic acid metabolism.*

as heat, as is the case when free acetic acid is formed as the product. If acetyl CoA is hydrolyzed to form CoA and acetic acid, large amounts of heat are given off.

The structures of the coenzymes involved in pyruvic acid metabolism are shown in Fig. 8-2. Note that sulfur is an important element in the structure of all three and that the B vitamin, pantothenic acid, is part of the molecule of coenzyme A. The isolation and identification of coenzyme A was one of the major advances of modern biochemistry. Not only is coenzyme A of intrinsic interest as an active biochemical reagent, but it is essential for many diverse reactions. Although the exact mechanisms of the series of reactions outlined in Fig. 8-1 are not clearly understood at the present time, the formation of the main product (acetyl CoA) is well established. Acetyl CoA is the key product of the metabolism of a number of organic acids. Additional metabolism that takes place, through oxidation of this C_2 unit in the citric acid cycle, is linked to reactions that transfer electrons ultimately to molecular oxygen. Consequently, in the presence of oxygen the reactions go to completion and liberate a large amount of energy that is conserved as ATP.

The citric acid cycle is initiated by the condensation of the active acetic acid with oxalacetic acid to form citric acid, a six-carbon compound containing three carboxyl groups (Fig. 8-3). Note that the energy of the acetyl CoA link is used in the condensation, and, as a result, the coenzyme A molecule is readily removed. In the ensuing reactions, citric acid is rearranged, by the removal and addition of water, to form isocitric acid.

As shown in Fig. 8-3, isocitric acid is then oxidized to form a new keto acid which rapidly loses CO_2 to form α-ketoglutaric acid. Since

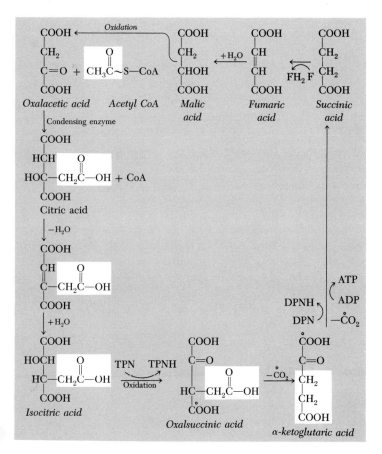

Figure 8-3 **Citric acid cycle.** (°Indicates source of carbon dioxide from the oxalsuccinic acid and α-ketoglutaric acid.)

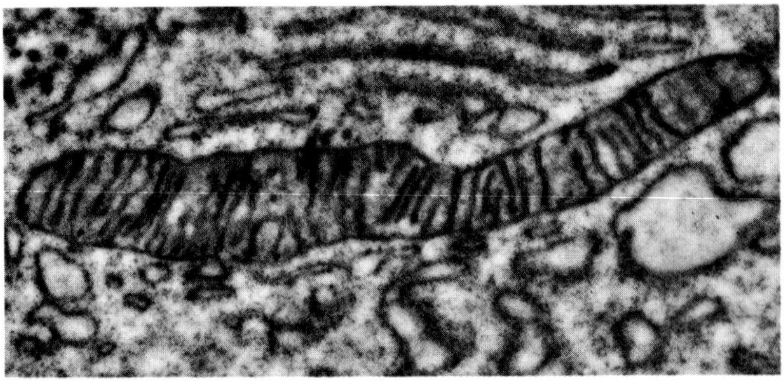

Figure 8-4 **Highly magnified electron micrograph** *of a mitochondrion in a mouse pancreatic cell. The outer boundary of the mitochondrion is a double structure, with the inner layer being continuous with the inner cross membranes.* [Courtesy of Dr. Bryce L. Munger.]

this latter keto acid is metabolized in a series of steps that are analogous to the initial steps in pyruvic acid metabolism, the oxidative decarboxylation of α-ketoglutaric acid results in the formation of succinic acid, CO_2, DPNH, and ATP. Succinic acid is then oxidized to fumaric acid, which, upon hydration, leads to the formation of malic acid. The oxidation of malic acid regenerates the oxalacetic acid and thus closes the cycle by entering the initial reaction, the condensation of acetyl CoA to form citric acid. When all the necessary enzymes are present, this cycle can be initiated by the addition of any of the compounds within the cycle. To keep the cycle going, however, acetyl CoA must be supplied.

This integrated group of enzymes—the dehydrogenases, the decarboxylases, and so forth—which engineer the complex series of reactions of the Krebs cycle, are localized in the mitochondria (Fig. 8-4) of the cell. Bound to this structure in some unknown way are the various coenzymes, which include coenzyme A, DPN, the flavins, and the cytochromes as well as a new electron-transport coenzyme called Q. This latter is a quinone and is related to vitamin K. Coenzyme Q is essential for the oxidation of reduced flavins. The mitochondria from several different kinds of cells have been isolated, and have been shown to be capable of carrying on the final oxidative stages of cell metabolism by themselves. Of considerable significance is the fact that the reduced coenzymes (DPNH or TPNH) can be oxidized by the mitochondria with the liberation of energy. This *oxidative phosphorylation,* which is known to require the other coenzymes, is one of the crucial events in oxidative metabolism. The energy liberated in the oxidation of these cofactors is used to synthesize ATP.

The electron flow from the high energy level of reduced DPN to the lower energy level of the other coenzymes thus liberates energy to form ATP molecules. Recent investigations on mitochondrial fragments indicate that there are three sites for the formation of ATP during the oxidation of DPNH by oxygen. These three reaction sites can be indicated as follows where the arrows indicate direction of electron flow:

DPNH $\longrightarrow$ flavoprotein $\longrightarrow$ coenzyme Q   one ATP formed
coenzyme Q $\longrightarrow$ cytochrome b $\longrightarrow$ cytochrome c one ATP formed
cytochrome c $\longrightarrow$ cytochrome oxidase $\longrightarrow$ O_2 one ATP formed

Thus, the oxidation of one mole of DPNH by the mitochondrial enzyme complex leads to the synthesis of three moles of ATP. Consequently, the transfer of two electrons (2H) to oxygen gives rise to the esterification of three inorganic phosphate molecules. Since one oxygen atom is used (2H + O $\rightarrow$ H_2O), the ratio of the phosphate taken up to oxygen used is 3 (P/O = 3). Note that in the oxidation of one pyruvic acid molecule, four molecules of reduced pyridine nucleotide are formed, and the oxidation of them leads to the synthesis of 12 ATP molecules. Two ATP molecules are obtained in the oxidation of succinic acid and one in the oxidation of ketoglutaric acid. The total combustion of pyruvic acid to CO_2 and H_2O thus leads to the net synthesis of 15 ATP molecules. It is no wonder that the mitochondria have been called the powerhouses of the cell. Oxidative phosphorylation is one of the most interesting and intriguing problems of modern biochemistry and much work is being done to determine the mechanisms involved in generating ATP in oxidative processes of this type.

Although the citric acid cycle is most often considered in terms of the oxidation of a substrate and the generation of ATP, it is also valuable in the production and utilization of the carbon skeletons of many other compounds. For many years, the breakdown of carbohydrates was thought to be primarily a degradation process whose sole purpose was the liberation of energy. This is true as far as energy production is concerned, but we now know that the formation of key intermediates in the breakdown of carbohydrates is very significant in the synthesis of compounds of biological interest. Indeed, it has been proposed that during the rapid growth of the cell, the principal function of the citric acid cycle is to supply the cellular carbon skeletons for biosynthesis.

It is also significant that during evolution various cell types have not developed different structures or systems for the metabolism of sugar fermentation products and other cellular foods. The mitochondrion, for example, also functions as the cellular furnace for the combustion of fatty acids, amino acids, and other fuels.

Proteins, carbohydrates, and lipids constitute the bulk of organic matter in cells. Lipids, or fats, are usually very insoluble in water, and as a consequence such organic solvents as ethanol or ether must be used to extract them from the cell. The lipids are a heterogeneous group of chemicals. The simplest ones contain only carbon, hydrogen, and oxygen, and by hydrolysis they yield glycerol and fatty acids. Most fatty acids are long straight chains in which the carbon atoms contain either the maximum number of hydrogen atoms (saturated) or a fewer number of hydrogens (unsaturated). The fatty acids may also have branched chains. A few examples are listed below; all contain the carboxyl (acid) group.

$CH_3CH_2CH_2COOH$—butyric acid—found in butter
$CH_3(CH_2)_6COOH$—octanoic acid—found in coconut oil
$CH_3CH{=}CH\ COOH$—crotonic acid—found in croton oil
$CH_3(CH_2)_7{-}CH{=}CH(CH_2)_7COOH$—oleic acid—found in animal and plant fats

The fats are a concentrated food source, since they supply more than twice as many calories per gram as do carbohydrates and proteins. Cells are able to synthesize most, if not all, the fatty acids from the carbon skeleton of carbohydrates in the diet, and the few fatty acids that they cannot synthesize must be taken in with the diet, i.e., are essential. The amount of essential fatty acids required is small, and is provided by almost any diet. As we shall consider in greater detail later, the complex fats that are synthesized by cells are necessary not only because of their energy source, but because they form structural components of the cell. In particular, the cell membrane, as well as other submicroscopic particles in the cell that contain membranes, has a fat or lipid structure.

The synthesis of a triglyceride occurs primarily in liver and in adipose tissue and is shown in Fig. 8-5. Glycerophosphoric acid (or dihydroxyacetone phosphate) reacts with the coenzyme A derivatives of fatty acids to form phosphatidic acid. The reaction proceeds at a high rate when saturated and unsaturated C_{16} and C_{18} fatty acid CoA derivatives are present.

The removal of the phosphate group from phosphatidic acid by a specific phosphatase yields a 1,2-diglyceride which in turn reacts with another fatty acid CoA to form a neutral triglyceride.

In animals free fatty acids are quite low in concentration. They are usually present in the form of an ester. Free fatty acids can be

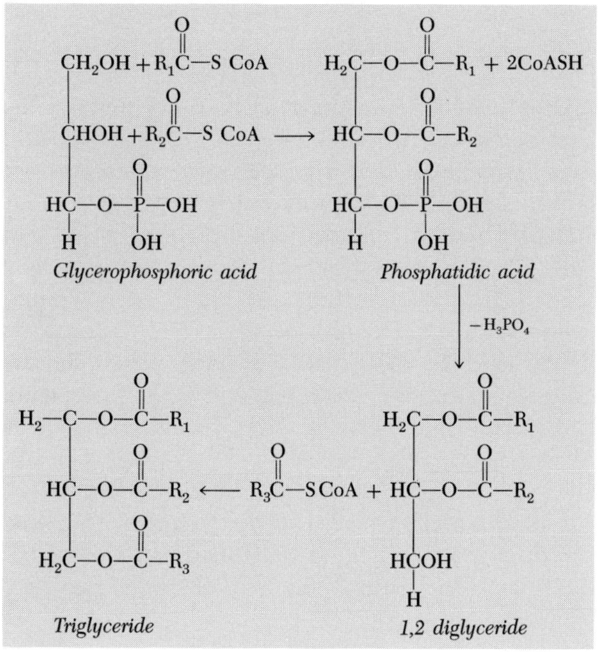

$$CH_2OH + R_1\overset{\displaystyle O}{\overset{\|}{C}}\!-\!S\,CoA \qquad H_2C\!-\!O\!-\!\overset{\displaystyle O}{\overset{\|}{C}}\!-\!R_1 + 2CoASH$$

$$CHOH + R_2\overset{\displaystyle O}{\overset{\|}{C}}\!-\!S\,CoA \longrightarrow HC\!-\!O\!-\!\overset{\displaystyle O}{\overset{\|}{C}}\!-\!R_2$$

$$HC\!-\!O\!-\!\overset{\displaystyle O}{\overset{\|}{P}}\!-\!OH \qquad HC\!-\!O\!-\!\overset{\displaystyle O}{\overset{\|}{P}}\!-\!OH$$

$$H \quad OH \qquad\qquad H \quad OH$$

Glycerophosphoric acid   *Phosphatidic acid*

$$-H_3PO_4$$

$$H_2\!-\!\overset{\displaystyle O}{\overset{\|}{C}}\!-\!O\!-\!C\!-\!R_1 \qquad H_2C\!-\!O\!-\!\overset{\displaystyle O}{\overset{\|}{C}}\!-\!R_1$$

$$HC\!-\!O\!-\!\overset{\displaystyle O}{\overset{\|}{C}}\!-\!R_2 \longleftarrow R_3\overset{\displaystyle O}{\overset{\|}{C}}\!-\!S\,CoA + HC\!-\!O\!-\!\overset{\displaystyle O}{\overset{\|}{C}}\!-\!R_2$$

$$H_2C\!-\!O\!-\!\overset{\displaystyle O}{\overset{\|}{C}}\!-\!R_3 \qquad HCOH$$

$$H$$

Triglyceride   *1,2 diglyceride*

Figure 8-5   **The synthesis of a triglyceride.**

readily converted into the CoA derivative by the following two reactions that are catalyzed by a single enzyme.

$$R\overset{\displaystyle O}{\overset{\|}{C}}\!-\!OH + ATP \rightleftharpoons R\overset{\displaystyle O}{\overset{\|}{C}}\!-\!AMP + PP$$

$$R\overset{\displaystyle O}{\overset{\|}{C}}\!-\!AMP + CoASH \rightleftharpoons R\overset{\displaystyle O}{\overset{\|}{C}}\!-\!S\!-\!CoA + AMP$$

The first step involves the activation of the carboxyl group of the fatty acid by the transfer of adenylic acid (AMP) from ATP and liberation of inorganic pyrophosphate (PP). The activated carboxyl reacts with the SH group of coenzyme A to form the thiolacyl intermediate and free AMP.

In contrast to the simple lipids are the waxes and related substances in which the glycerol is replaced by a longer-chain alcohol. Beeswax, for instance, is an ester of palmitic acid (a 16-carbon saturated fatty acid) and myricyl alcohol (a 30-carbon-chain saturated alcohol). In the simple lipid, one of the fatty acids may be replaced by compounds containing phosphorus and nitrogen to form the phosphatides, *lecithin* and *cephalin*, which frequently represent the major portion of cellular lipids. These compounds are soluble in both water and fats and therefore serve a vital role in the cell by binding water-soluble compounds (i.e., proteins) and lipid-soluble compounds together. Lecithin is a key structural material in the cell membrane, because it can maintain continuity between the aqueous and lipid phases of the inside and outside of the cell. In addition we know that the function of certain enzymes depends upon their attachment to a lipid such as lecithin.

FATTY ACID OXIDATION

The oxidation of fatty acids proceeds in several distinct steps, but the final product is the active acetic acid unit, acetyl CoA. Fatty acids, then, feed two carbon units into the citric acid cycle in the same way that pyruvic acid does. One crucial oxidative step in fatty acid metabolism we must now consider. The fatty acids of biological importance include a large series of straight-chain acids that begins with formic acid and continues through acetic acid to compounds with more than 20 carbon atoms. Interestingly enough, the naturally occurring fatty acids are predominantly even-numbered.

Let us examine the metabolism of the four-carbon butyric acid. The first thing discovered about fatty acid oxidation was that the mitochondria contain the necessary enzymes and cofactors for this process. It was next observed that a small amount of ATP is essential to initiate fatty acid oxidation, and subsequent developments have demonstrated that this activation by ATP is required for the formation of the coenzyme A derivative of fatty acids. The carboxyl group of the fatty acids, such as butyric, reacts with ATP to form an energy-rich adenylic acid derivative. This intermediate then reacts with coenzyme A to form butyryl CoA.

As indicated in Fig. 8-6, the coenzyme A derivative is oxidized between the number 2 and 3 carbons to form a double bond. This is called *β-oxidation*. The electron acceptor in this oxidation is usually a flavin. Water is now added across this double bond to form an OH group on the beta carbon, a process that is analogous to the hydration of fumaric acid to form malic acid. Oxidation now occurs, leading to the formation of a double-bond oxygen on carbon number 3 that is capable of reacting with coenzyme A to split off two coenzyme A units. At present, the cycling of CoA in the breakdown of fatty acids through β-oxidation is probably a reasonable explanation for the metabolism of fatty acids of any chain length. Thus, if we start with an even-numbered fatty acid, we will obtain an even number of acetyl CoA units. A C_8 fatty acid, therefore, gives rise sequentially to: $C_8CoA \rightarrow C_6CoA \rightarrow C_4CoA \rightarrow 2C_2CoA$. In each step, a C_2CoA is generated.

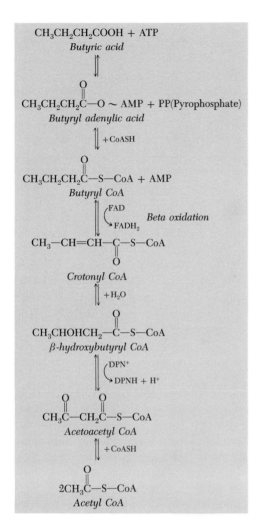

$$CH_3CH_2CH_2COOH + ATP$$
Butyric acid

$$CH_3CH_2CH_2\overset{\displaystyle O}{\overset{\|}{C}}-O \sim AMP + PP(\text{Pyrophosphate})$$
Butyryl adenylic acid

$+ CoASH$

$$CH_3CH_2CH_2\overset{\displaystyle O}{\overset{\|}{C}}-S-CoA + AMP$$
Butyryl CoA

FAD
FADH₂ *Beta oxidation*

$$CH_3-CH=CH-\underset{\underset{\displaystyle O}{\|}}{C}-S-CoA$$
Crotonyl CoA

$+ H_2O$

$$CH_3CHOHCH_2-\overset{\displaystyle O}{\overset{\|}{C}}-S-CoA$$
β-hydroxybutyryl CoA

DPN⁺
DPNH + H⁺

$$CH_3\overset{\displaystyle O}{\overset{\|}{C}}-CH_2\overset{\displaystyle O}{\overset{\|}{C}}-S-CoA$$
Acetoacetyl CoA

$+ CoASH$

$$2CH_3\overset{\displaystyle O}{\overset{\|}{C}}-S-CoA$$
Acetyl CoA

Figure 8-6   Fatty acid metabolism.

The metabolism of odd-numbered fatty acids presents some special problems. If we start with a C_9 fatty acid, the steps appear to be the same ($C_9 \rightarrow C_7 \rightarrow C_5 \rightarrow C_3$) until we get to C_3CoA (propionyl CoA). Present evidence indicates that this active form of propionic acid must take up CO_2 to form the even-carbon succinic acid before it can be further metabolized. This CO_2 fixation is rather complicated and not yet well understood. Although active C_2 units (acetyl CoA) are used in the synthesis of fatty acids, the pathway is apparently not a simple reversal of the scheme that has been illustrated in Fig. 8-6.

Note that in the oxidation of butyric acid we obtain two acetyl

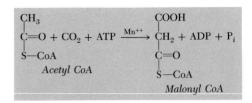

CoA units, which, on entering the citric acid cycle, lead to the generation of 24 ATP molecules. In addition, five other ATP molecules are generated during the initial oxidative reactions that give rise to the formation of acetyl CoA. It is evident, therefore, that much more energy can be obtained from the oxidation of fatty acids than from the equivalent carbon-length of a carbohydrate.

The *de novo* synthesis of fatty acids involves enzymes and cofactors which are different from those used in degradation, although acetyl CoA again functions as a major carbon source for lipogenesis. Furthermore, the synthesis occurs in the cytoplasm in a complex that contains all the lipogenic enzymes in a unit that has a particle weight of well over two million. The two key carbon sources for fatty acid synthesis are acetyl CoA and malonyl CoA. The latter is formed from acetyl CoA by a CO_2 fixation reaction shown in Fig. 8-7. The enzyme (acetyl CoA carboxylase) contains the B vitamin, biotin, to which the CO_2 is first added and then transferred to acetyl CoA.

A second important enzyme that occurs in the complex is called acyl carrier protein (ACP). It binds acyl intermediates during the formation of long-chain fatty acids. The active group on the protein is an SH which acts very much like the SH in CoA in accepting acetyl and malonyl groups. As shown in Fig. 8-8, the acetyl CoA and malonyl CoA react with the Sh group in ACP to form the key intermediates of fatty acid synthesis. The initial reaction is the only step in which free acetyl CoA is used directly. Subsequently, malonyl-S-ACP serves the primary carbon donor for chain elongation. In step 3 acetyl-S-ACP reacts with malonyl-S-ACP to form acetoacetyl-S-ACP with the release of the CO_2. The reduction of acetoacetyl-S-ACP to the butyryl derivative requires TPNH rather than DPNH, the reduced pyridine nucleotide formed in fatty acid degradation. A second malonyl-S-ACP reacts with the butyryl intermediate to form a six-carbon intermediate which is in turn reduced by the types of reactions shown in 4 through 6. The process continues using malonyl-S-ACP until palmitic acid is formed. Note that the methyl group of the initial acetyl-S-ACP remains at the end of the growing chain.

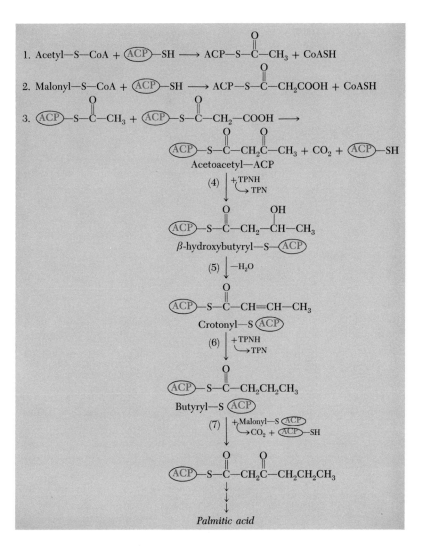

1. Acetyl—S—CoA + (ACP)—SH $\longrightarrow$ ACP—S—$\overset{\overset{\displaystyle O}{\|}}{C}$—CH$_3$ + CoASH

2. Malonyl—S—CoA + (ACP)—SH $\longrightarrow$ ACP—S—$\overset{\overset{\displaystyle O}{\|}}{C}$—CH$_2$COOH + CoASH

3. (ACP)—S—$\overset{\overset{\displaystyle O}{\|}}{C}$—CH$_3$ + (ACP)—S—$\overset{\overset{\displaystyle O}{\|}}{C}$—CH$_2$—COOH $\longrightarrow$

(ACP)—S—$\overset{\overset{\displaystyle O}{\|}}{C}$—CH$_2\overset{\overset{\displaystyle O}{\|}}{C}$—CH$_3$ + CO$_2$ + (ACP)—SH

Acetoacetyl—ACP

(4) $\quad$ +TPNH $\searrow$ TPN

(ACP)—S—$\overset{\overset{\displaystyle O}{\|}}{C}$—CH$_2$—$\overset{\overset{\displaystyle OH}{|}}{C}$H—CH$_3$

β-hydroxybutyryl—S—(ACP)

(5) $\quad$ —H$_2$O

(ACP)—S—$\overset{\overset{\displaystyle O}{\|}}{C}$—CH=CH—CH$_3$

Crotonyl—S (ACP)

(6) $\quad$ +TPNH $\searrow$ TPN

(ACP)—S—$\overset{\overset{\displaystyle O}{\|}}{C}$—CH$_2CH_2CH_3$

Butyryl—S (ACP)

(7) $\quad$ +Malonyl—S (ACP) $\searrow$ CO$_2$ + (ACP)—SH

(ACP)—S—$\overset{\overset{\displaystyle O}{\|}}{C}$—CH$_2\overset{\overset{\displaystyle O}{\|}}{C}$—CH$_2CH_2CH_3$

Palmitic acid

Figure 8-8   *The synthesis of a fatty acid.*

In addition to the cytoplasmic enzyme complex for *de novo* synthesis of fatty acids there occurs in the mitochondria an enzyme system that catalyzes the elongation of existing fatty acids by the addition of two carbon units derived from acetyl CoA.

Since acetyl CoA carboxylase is the rate-limiting step in fatty acid synthesis in mammals, it is one of the major sites for control. For example, the enzyme is greatly stimulated by intermediates of the citric acid cycle which may be expected to be in high concentration on a high carbohydrate diet. In addition, a source of TPNH is essential and significant amounts can be obtained from the dehydrogenation reactions of the phosphogluconic acid oxidative pathway.

ENZYME COMPLEXES

The enzymes used in fatty acid synthesis are associated as a complex in the cell. Attempts to separate the complex into distinct enzymes have been unsuccessful. It would appear that the separated enzymes are inactive, perhaps because they do not have the proper shape, but when they are aggregated into a *multienzyme complex* they can carry out all the necessary reactions in sequence. There are obvious advantages to such a system. If we are dealing with a chemical series, and the product of one reaction is the substrate for the next enzyme, the enzyme complex acts as a cellular assembly line. The complex would be considerably more efficient than a system in which each of the enzymes and substrates are separately in solution and haphazardly come together to react. Also, control is more effectively exercised because blocking one reaction of the series blocks the entire system. The mitochondrion and the chloroplasts in the cell are other examples of multienzyme complexes.

AMINO ACID METABOLISM

The proteins occupy a central place in both the structural and the dynamic aspects of living matter. Along with the nucleic acids, they compose the most important macromolecular structures of cells. Proteins are composed of amino acids, and it is partly the difference in amino acid composition that gives the proteins some of their unique properties. A supply of amino acids, either from the diet or from the biosynthetic machine, is obviously essential to cells and tissues for the synthesis of proteins. When these essential amino acids are acquired in the diet, the remainder of the nitrogen needed for protein synthesis can be supplied in the form of ammonium salts. Ammonia is an intermediate in nitrogen metabolism, and most organisms, when given an adequate amount of utilizable carbon compounds and other essential growth elements, can readily employ ammonia as their principal source of protein nitrogen. A crucial reaction in the uptake and incorporation of ammonia into proteins involves one of the intermediates in the citric acid cycle. As outlined in Fig. 8-9, α-ketoglutaric acid can be converted into the amino acid glutamic acid by a process called *reductive amination*, in which TPNH is the reducing power and ammonia is the nitrogen source. This reaction is reversible, and, in fact, the enzyme that

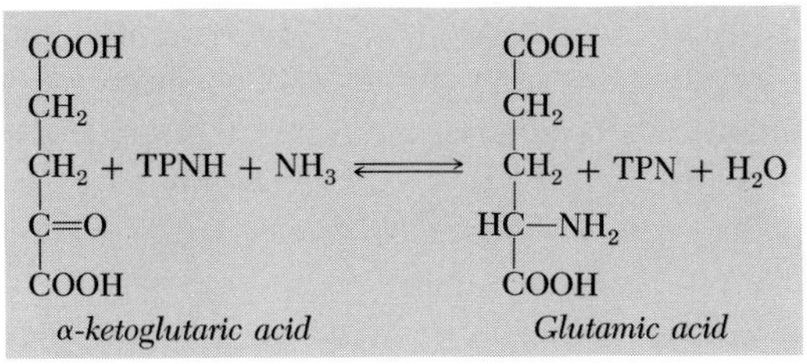

COOH   COOH
| |
CH$_2$ CH$_2$
| |
CH$_2$ + TPNH + NH$_3$ $\rightleftharpoons$ CH$_2$ + TPN + H$_2$O
| |
C=O   HC—NH$_2$
| |
COOH   COOH
α-ketoglutaric acid   Glutamic acid

Figure 8-9   *Synthesis of glutamic acid.*

catalyzes the reaction is called glutamic dehydrogenase. It is clear that this reaction serves as a main link between amino acid and carbohydrate metabolism.

Once ammonia is converted into amino nitrogen, it can be transferred to other carbon skeletons to form a different amino acid. This process of *transamination,* which, as shown in Fig. 8-10, involves an amino acid and a keto acid. Since the reaction is reversible, the process represents one of the principal metabolic pathways for the formation and deamination of amino acids. Enzymes called transaminases, which catalyze these various reactions, are known to occur in a variety of animal tissues, plants, and microorganisms. The B vitamin, pyridoxine, in the form of pyridoxal phosphate, is an essential cofactor for all transaminases. Two of the intermediates in the citric acid cycle, therefore, serve as carbon skeletons for the synthesis of two amino acids. The process of reductive amination and transamination can lead, as

Figure 8-10   *Transamination.*

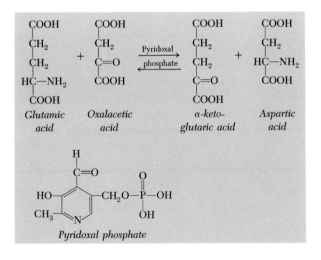

far as we know, to the incorporation of ammonia into a large number of carbon skeletons that are formed from carbohydrate and fatty acid metabolism. How particular carbon skeletons are formed, then, is the major problem in amino acid biosynthesis, and, for the most part, it appears that the intermediates of the citric acid cycle are the initial compounds for the formation of amino nitrogen.

During evolution animals have lost the ability to make the carbon chain for certain amino acids. These amino acids must be supplied in the diets and are therefore called essential amino acids. The essential amino acids for rats are arginine, histidine, isoleucine, leucine, lysine, methionine, phenylalanine, threonine, tryptophan, and valine. Although equally important, the following amino acids can be synthesized by mammals and therefore are called nonessential: alanine, aspartic acid, cystine, glutamic acid, glycine, hydroxyproline, proline, serine, tyrosine.

Looking at the list of the essential amino acids it becomes clear why the recently developed strain of corn which produces a high lysine content is such an important development. In most strains, about 50 per cent of the protein made in corn is *zein*, which is very low in lysine. Thus, if man's only source of amino acids were corn, he would develop a lysine deficiency and consequently could not make proteins (this condition is known as *kwashiorkor* and occurs in economically undeveloped areas, primarily in the tropical regions of the world). The new strain of corn (opaque 2—a simple single gene difference) is changed in such a way that much less zein is made and the concentration of other proteins that contain more lysine is increased. Opaque 2 should thus provide an important source of nutrition to man.

The relationship between the citric acid cycle and amino acid metabolism is clearly shown in the series of reactions that lead to the synthesis of the amino acid arginine, and to the major nitrogenous excretion product of man, urea. When excess amino acids are fed to animals, much of the excess nitrogen is excreted as urea. The unraveling of the mechanisms involved in urea synthesis was another milestone in our understanding of cyclic processes in biochemistry. As indicated below, the "ornithine cycle" regulates the removal of ammonium ions and depends for its functioning on the presence of the citric acid cycle. The essential compounds that are required to maintain the continued synthesis of urea are the following: ATP, aspartic acid, ammonia, and CO_2.

As shown in Fig. 8-11, glutamic acid can be converted into ornithine by reduction and transamination. In the presence of CO_2, ATP, and ammonia, ornithine is converted into citrulline by the addition of NH_2COOH, which is made from CO_2 and NH_3 in the presence of

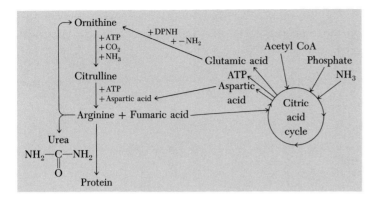

Figure 8-11   *Urea and arginine synthesis.*

ATP and the appropriate enzyme. The true intermediate is carbamyl phosphate:

$$CO_2 + NH_3 + ATP \longrightarrow NH_2\overset{\overset{O}{\|}}{C}O \sim \overset{\overset{O}{\|}}{\underset{\underset{OH}{|}}{P}}-OH + ATP$$

It has been shown recently that the vitamin biotin is an essential cofactor for the enzyme that catalyzes this reaction. As noted previously, biotin is essential in other reactions involving CO_2 fixation.

Aspartic acid adds a $-NH_2$ group to citrulline to form the amino acid arginine. In the liver, if arginine is not used for protein synthesis, it is broken down by the enzyme arginase to form urea and ornithine. The ornithine or urea cycle thus leads to the excretion of two nitrogen atoms in the form of urea, and, as is evident from the scheme, the process depends on the immediate participation of the citric acid cycle for a supply of ATP and of the appropriate carbon-skeleton intermediates for the synthesis of aspartic and glutamic acids.

The importance of carbohydrate metabolism in the synthesis of amino acids is further emphasized in the mechanism of synthesis of the sulfur-containing amino acid cysteine. As illustrated in Fig. 8-12, the initial step involves the transfer of an active two-carbon unit from acetyl CoA to serine to form the intermediate O-acetyl serine plus free coenzyme A. The enzyme that catalyzes this reaction (serine transacetylase) is inhibited by cysteine; consequently, if cysteine is available from food, the synthetic machinery is inactive and thus cellular intermediates are preserved for other essential functions. This so-called feedback inhibition is an important mechanism for the regulation of cellular metabolism and will be discussed in detail later.

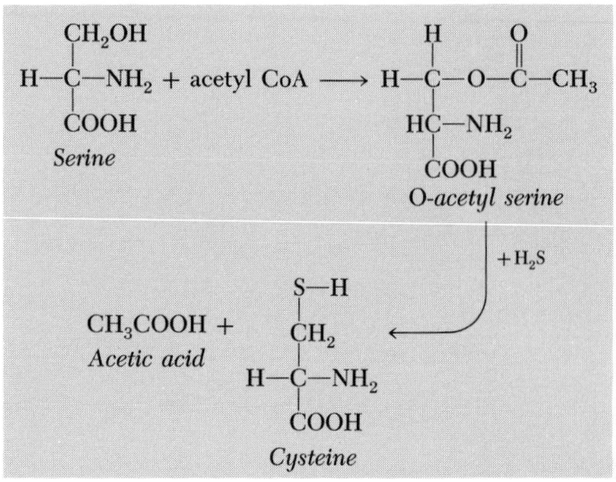

Figure 8-12 *Cysteine biosynthesis* in plants and microorganisms.

The O-acetyl serine formed in the first step reacts with hydrogen sulfide (H_2S) to form cysteine and acetic acid. The synthesis of the enzyme that catalyzes this reaction is prevented by the presence of high concentrations of cysteine. This repression of enzyme synthesis is another important mechanism for the regulation of metabolism and will be discussed in a subsequent chapter.

This mechanism of cysteine biosynthesis has been shown to occur in microorganisms and certain plants. In mammals the cysteine is made from the essential amino acid, methionine. As shown in Fig. 8-13, the

Figure 8-13 *Cysteine biosynthesis in mammals from methionine.*

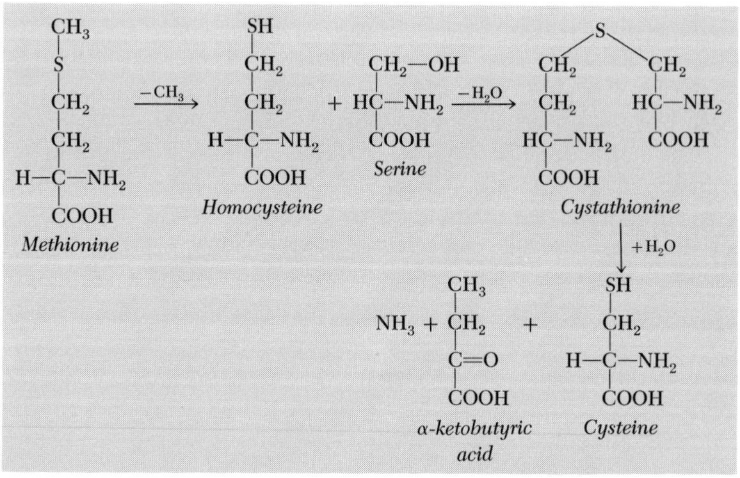

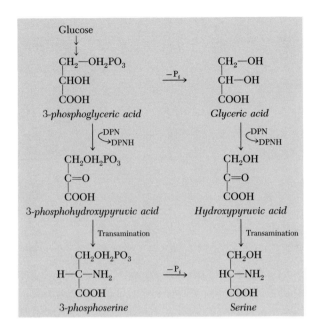

Figure 8-14   *Serine biosynthesis.*

initial step involves the removal of a methyl group to form homocysteine. Serine is used in the next step to condense with homocysteine to form cystathionine. The latter is now cleaved to form cysteine, α-ketobutyric acid, and ammonia. The net effect of this series of reactions is the transfer of an SH group (transsulfuration) from homocysteine to serine.

It should be pointed out that the synthesis of methionine (an essential amino acid for man) in plants and microorganism makes use of cysteine. Cystathionine is the key intermediate.

It is evident that serine is an important amino acid for the synthesis of other amino acids. The pathway of biosynthesis is shown in Fig. 8-14; 3-phosphoglyceric acid provides the primary carbon chain. Two pathways have been established, but the one that utilizes 3-phosphoserine is probably the major one. As the scheme indicates, the major steps involve an oxidation using DPN as the electron acceptor and a subsequent transamination of either 3-phosphohydroxypyruvic, using glutamic acid as the amino donor, or hydroxypyruvic acid, using alanine as the amino donor. Hydrolysis of phosphoserine by a specific phosphatase yields serine.

Serine is also converted into the amino acid glycine (NH_2—CH_2—$COOH$). This reaction is significant in that it provides the prime

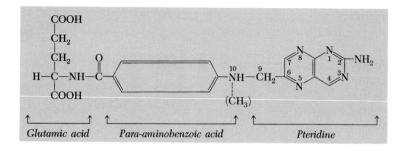

Figure 8-15 *Structure of folic acid.*

Glutamic acid   Para-aminobenzoic acid   Pteridine

metabolic source of one-carbon units, i.e., $-CH_3$, $-CH_2OH$, and
$-CHO$. The transformation of serine to glycine and a C_1 unit is
fairly complicated involving the important cofactor *folic* acid whose
structure is shown in Fig. 8-15. Folic acid contains the vitamin *para*-
aminobenzoic acid as well as a pteridine group and glutamic acid.
A reduced form of folic acid is the active form which serves as the
initial acceptor of the $-CH_2OH$ group from serine; the other product
of this reaction is glycine. Through a series of transformations the
$-CH_2OH$ group (bound at the N10 nitrogen of the folic acid) is
converted into a methyl group that can now participate in trans-
methylation reactions such as in the conversion of homocysteine to
methionine.

In microorganisms formic acid can act as the initial one-carbon
source. In this case ATP must be used also, and the initial folic acid
compound formed has an N10 aldehyde group ($-CHO$). In a series
of steps the $-CHO$ is reduced by TPNH to $-CH_2OH$ and finally to
$-CH_3$.

Glycine, the simplest of all the amino acids, is essential for a number
of key biosynthetic reactions. It condenses with succinyl COA to form
the initial precursor to pyrrol, the major precursor to porphyrins (found
in hemoglobin, cytochromes, chlorophyll). Glycine is also essential for
the synthesis of creatine, uric acid, and hippuric acid.

The metabolism of the amino acid, tryptophan, in animals and
plants is of considerable interest. As indicated in Fig. 8-16, it is the
primary source of the plant growth hormone, auxin. In man, tryptophan can be converted into 5-hydroxytryptamine (serotonin). Serotonin
is present in the blood platelets and is released when there is damage
to the blood vessels. It acts as a very potent vasoconstrictor and therefore may be of some importance in preventing blood loss before clot-
ting occurs. Since serotonin is concentrated rather highly in the brain,

its role as a possible neurohumoral agent has been suggested. Certain drugs that have a profound effect on the activities of the central nervous system are potent antagonists of serotonin. Reserpine, found in the Indian plant *Rauwolfia serpentina*, when administered to man considerably decreases the activity of the central nervous system and at the same time stimulates the excretion of serotonin in the urine.

In most animals, plants, and microorganisms tryptophan can be converted directly to niacin ribonucleotide, the immediate precursor to DPN.

The story of the biosynthesis of aromatic amino acids in comparative biochemistry is intriguing; the interested student will find it rewarding to consult advanced biochemistry texts on this subject.

The metabolism of phenylalanine is of particular interest because it is known to be implicated in an inherited metabolic disease. In normal individuals the phenylalanine that is not used in protein synthesis is hydroxylated to form the amino acid tyrosine. This reaction

Figure 8-16   **Formation of auxin and serotonin** *from tryptophan.*

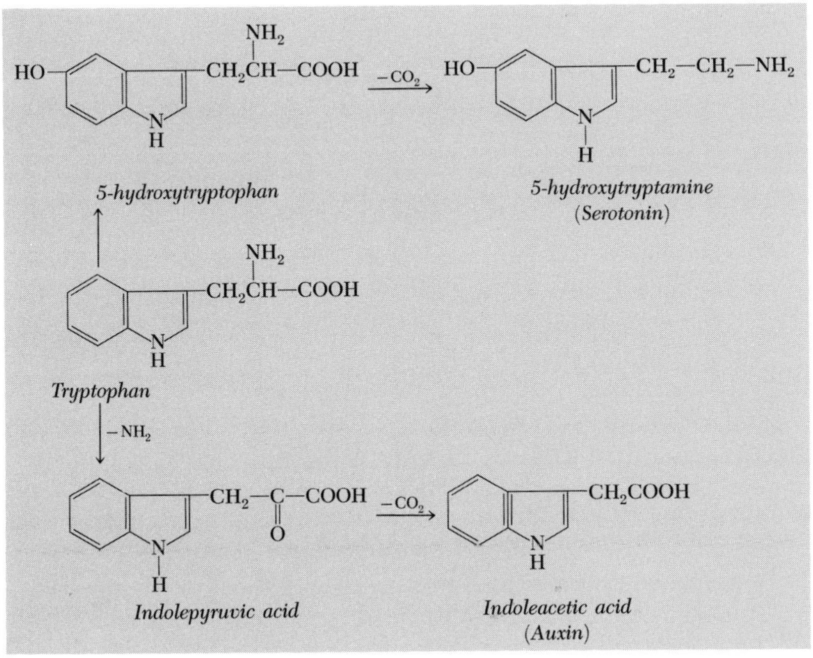

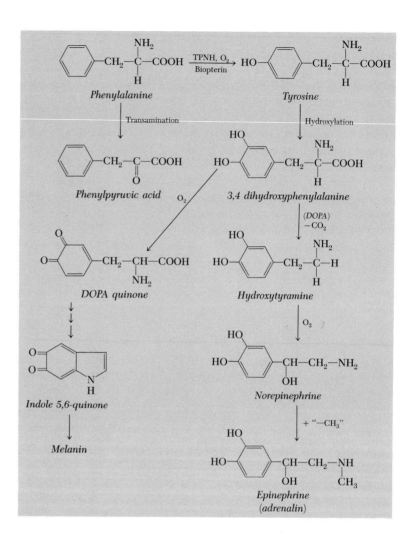

*Figure 8-17   Metabolism of phenyl-
alanine and tyrosine.*

is shown in Fig. 8-17. The occurrence of a genetic defect (homozygous recessive genes) in about one in every 200 individuals leads to a loss of the enzyme (phenylalanine hydroxylase) that converts phenylalanine to tyrosine. In the absence of this enzyme excess phenylalanine is metabolized by way of phenylpyruvic acid, which is excreted in large amounts in the urine. This condition, known as phenylketonuria (PKU), causes mental retardation. It can be detected in infancy and by feeding tyrosine and keeping the phenylalanine low in the diet one can prevent, in part, the mental retardation. Little is known about how to prevent the occurrence of this inherited metabolic defect, and clearly this is an important area for future research.

As indicated in Fig. 8-17, tyrosine is the precursor to the adrenal hormone, epinephrine and to the dark pigment, melanin.

The mechanism of synthesis and degradation of a number of the amino acids is a fascinating subject, but it is not necessary for us to go into the details of all of these reactions at this time. The principles outlined for the synthesis and breakdown of the amino acids discussed hold for the metabolism of other amino acids. To pursue the many other interesting metabolic products of amino acid metabolism, some of which are extremely important to cellular function, you should consult a more advanced book on biochemistry.

The glycolytic, or fermentative, pathway of glucose metabolism represents the major route for the formation of pyruvic acid from carbohydrates. There are a number of other alternate pathways of carbohydrate metabolism, however, the most prominent of which has been called the oxidative or pentose phosphate shunt. This pathway has the advantage of providing a means for the combustion of glucose to carbon dioxide without the participation of the citric acid cycle. In addition, the pathway leads to the formation of ribose, the five-carbon sugar we have shown to be important in the synthesis of a number of coenzymes, ATP, and other cellular components. As discussed in another book in this series,° this pathway of carbohydrate metabolism is also intimately concerned in the carbon dioxide uptake in the photosynthetic organisms.

The first step in the metabolism of glucose by the oxidative pathway that is different from any of the steps in the fermentative pathway involves glucose-6-phosphate. As shown in Fig. 8-18, glucose-6-phosphate is oxidized to the corresponding 6-phosphogluconic acid. The electron acceptor in this case is TPN instead of DPN. Following the initial oxidation, 6-phosphogluconic acid is oxidized by TPN to form an intermediate which (see Fig. 8-18) is presumed to be 6-phospho-3-ketogluconic acid and is quickly decarboxylated to give the known products CO_2 and ribulose-5-phosphate. Ribulose-5-phosphate is readily converted into ribose-5-phosphate by the enzyme isomerase.

In the photosynthetic process, the primary acceptor of CO_2 is

° A. W. Galston, *The Life of the Green Plant*, 2nd ed. (Englewood Cliffs, N. J.: Prentice-Hall, 1964).

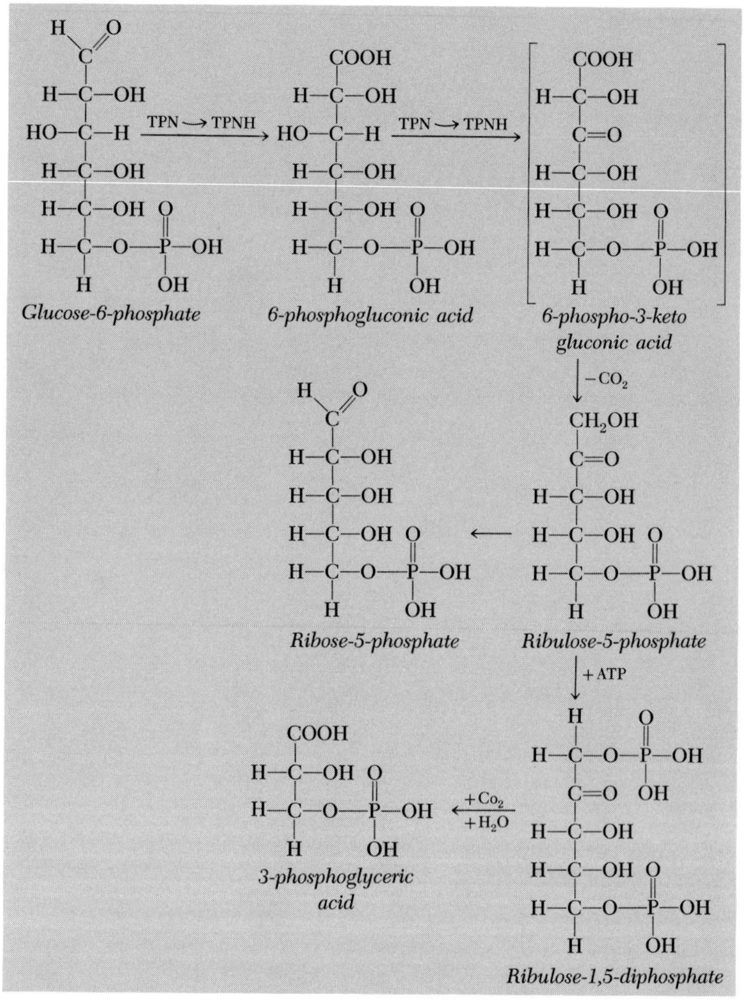

Figure 8-18   *Hexose monophosphate oxidation.*

apparently ribulose-1,5-diphosphate. The latter is formed from ribulose-5-phosphate and ATP. Photosynthesis will be discussed briefly in the next chapter.

From the last three chapters, we can see that great advances in the analysis of energy transformation and biosynthetic processes in organisms have been made during the past 30 years. The fundamental

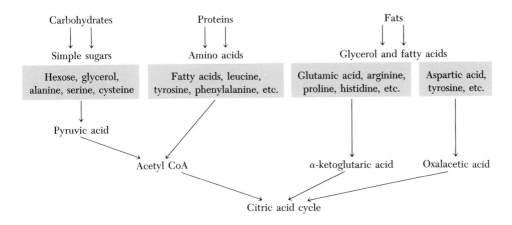

Figure 8-19 **Summary of the metabolism of fats,** *carbohydrates, and proteins.* [After Krebs.]

energy-liberating reactions that are necessary to maintain the diverse and sometimes complex cellular machines are superficially relatively simple. These reactions are examples of electron or hydrogen transport. To trigger biochemical reactions, we simply couple energy-yielding reactions to energy-consuming steps. The special form of chemical energy employed by biological systems for these coupling reactions is stored in the pyrophosphate bond of ATP.

As Krebs has recently emphasized, the first major stage of energy transformation in living matter terminates in the synthesis of ATP, at the expense of the free energy liberated during metabolism. In these energy transformations, we observe only a few basic processes, namely, oxidation, reduction, dehydration, hydration, decarboxylation, acetylation, phosphorylation, amination, and transamination. For carbohydrate, protein, and fat metabolism, the three key substances that are formed prior to any significant energy liberation are: acetyl CoA, α-ketoglutaric acid, and oxalacetic acid. These are broken down in the citric acid cycle to liberate over two-thirds of the cellular energy. The reactions of various food materials are summarized in Fig. 8-19.

9 SPECIAL PROBLEMS IN ENERGY TRANSFORMATION

LIGHT IS ABSORBED BY THE CHLOROPHYLL PIGMENTS IN PLANTS AND transformed by the cells into the chemical energy needed for the synthesis of carbohydrates. Priestley was the first to observe that green plants are capable of producing oxygen, although he did not know that light is essential in manufacturing it. During the next few years, it was clearly established that green plants could also remove the CO_2 produced during metabolism. Some time later the quantitative relationship between the CO_2 taken up and the oxygen produced was established. Plants, like animals, also respire, and we now know that they remove hydrogen from carbon skeletons and transfer them to oxygen, and also break carbon chains to form CO_2. These processes occur in the dark as well as in the light. Photosynthesis, therefore, must be the reverse of respiration; the light energy must be used to transfer hydrogen from water back to a carbon skeleton, which eventually comes from CO_2. Thus, in its simplest form, photosynthesis may be written as follows:

$$CO_2 + 2H_2O + light \longrightarrow (CH_2O) + O_2 + H_2O$$

To make a complete sugar molecule, six CO_2 molecules must be available. Therefore:

$$6CO_2 + 12H_2O + light \longrightarrow C_6H_{12}O_6 + 6O_2 + 6H_2O$$

If it seems peculiar that we put in water in the two equations and then take out water, the explanation lies in the fact that the resulting O_2 comes from the water and not from CO_2. In the photochemical process, water is presumed to split into a reducing part (H) and an oxidizing part (OH). The reaction of OH units is believed to give rise to some water and O_2.

The primary process in photosynthesis, therefore, is the absorption of light quanta by the chlorophyll molecule. Since chlorophyll absorbs in the red region of the spectrum (656 nm)—for this reason the leaves look green—we should suspect that the red quanta are the effective ones in photosynthesis. In general, this has been found to be true. Although we are not sure of the exact mechanism, it is clear that light energy must cause some of the electrons of the chlorophyll molecules to jump to an excited state. This excitation energy somehow splits water to form a highly reducing "H" and an oxidant "OH." Thus, in the light phase of photosynthesis, the photochemical decomposition of water produces a tremendous amount of potential energy.

Recent evidence indicates that there are probably two types of chlorophyll in green plants and that both must be excited by light in order for photosynthesis to proceed. One of these chlorophyll molecules absorbs light quanta in the red region of the spectrum and the other absorbs light quanta in the far-red region.

As shown in Fig. 9-1, when chlorophyll absorbs red light quanta it gives up an electron to cytochrome 550. The iron in the cytochrome accepts the electron. The oxidized chlorophyll reacts with water and becomes reduced, and in the process oxygen is liberated. Unfortunately, we do not understand the mechanism of this process, we do know, however, that at least four light quanta are required for each O_2 that is liberated. Thus, the overall reaction for the red light process can be described as follows:

$$4H_2O \longrightarrow 2H_2O + O_2 + 4H$$

The four hydrogens or electrons are used to reduce four chlorophyll molecules. The reduced cytochrome formed in the initial photochemical event is oxidized by another cytochrome at a lower energy level. When the electron is transported from cytochrome 550 to cytochrome b the potential energy release is conserved as phosphate bond energy and consequently ATP is formed.

The reduced cytochrome b is apparently complexed closely with a chlorophyll molecule because when this chlorophyll absorbs a red

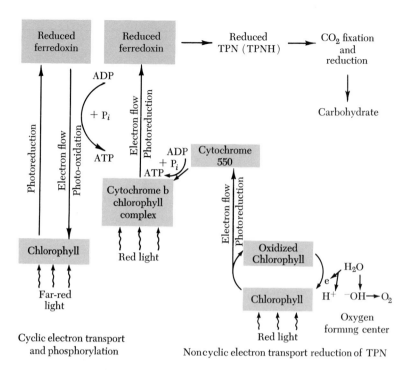

Figure 9-1 **Suggested scheme for electron transport** and phosphorylation during photosynthesis in green plants.

Cyclic electron transport and phosphorylation

Noncyclic electron transport reduction of TPN

light quantum the electrons in the reduced cytochrome b are raised to a high energy level and are now capable of reducing another iron-containing molecule called *ferredoxin* (see Fig. 9-1). The photo-oxidized cytochrome b is now available to accept additional electrons from reduced cytochrome 550.

The reduced ferredoxin passes its electrons on to TPN to form TPNH, the reducing potential needed to synthesize carbohydrate. Since two light quanta are needed to activate the phototransfer of one electron to ferredoxin, it is obvious that four quanta of light are needed to form one molecule of TPNH (2 electrons). Since we need four red light quanta to form one molecule of O_2 in the primary event, we can write a balanced equation for the formation of reduced TPN, which requires eight quanta as follows:

$$2TPN + 4HOH \xrightarrow{\text{(8 light quanta)}} 2TPNH + 2H_2O + O_2 + 2H^+$$

A second independent photochemical reaction occurs in the chloroplasts of green plants and makes use of *far-red* light. In this system the reduced ferredoxin formed is oxidized by the reduced chlorophyll and in the process ATP is formed (see Fig. 9-1). Since the electron transfer in this system is cyclic, the phosphorylation that occurs has been called *cyclic phosphorylation*. The phosphorylation occurring in both systems is known as *photophosphorylation*.

Reactions similar to the ones described above for green plants also occur in certain photosynthetic bacteria, except that the latter do not evolve oxygen. Instead of water they use other reducing agents, either inorganic or organic. The over-all equation for TPN reduction in photosynthetic bacteria can be summarized as follows:

$$2TPN + 2H_2A \longrightarrow 2TPNH + 2A + 2H^+$$

One of the primary roles of TPNH is to reduce CO_2 Before this can occur, CO_2 must be taken up by some reaction in the plant. Recent isotopic evidence indicates that the CO_2 adds to the 5-carbon sugar *ribulose diphosphate* which immediately splits into two molecules of *3-phosphoglyceric acid*. By reversing the glycolysis scheme and making use of reduced pyridine nucleotides and ATP, sugar can be synthesized. Another book in this series gives additional details on photosynthesis.[°]

Much remains to be done before we are absolutely sure that this is the correct scheme for photosynthesis. The photosynthetic apparatus is a highly structured system (see Fig. 9-2) and unravelling the machinery of this structure (quantosome) which carries out the various phases of photosynthesis is a challenging and exciting problem.

The oxidation of organic compounds during respiration, and during fermentation, gives off CO_2 to the air. In addition to this source, the burning of oil and coal also releases large amounts of CO_2 into the air. It is remarkable that the concentration of CO_2 within the air remains very nearly constant. Evidently the total rate of CO_2 production is exactly balanced by CO_2 photosynthetic consumption. The exact mechanism of the regulation of the atmospheric CO_2 by photosynthesis is not clearly understood. However, we do know that it is of great importance in affecting the Earth's temperature.

Light is essential not only for photosynthesis, but also for the production of chlorophyll. Chlorophyll begins as a substance called **protochlorophyll.** Protochlorophyll is different from chlorophyll in that it needs two additional hydrogens to make it into chlorophyll. Only in the presence of light can protochlorophyll change into chlorophyll. You can see that this is so by growing bean seedlings in the dark and

[°] A. W. Galston, *The Life of the Green Plant*, 2nd ed. (Englewood Cliffs, N. J.: Prentice-Hall, 1964).

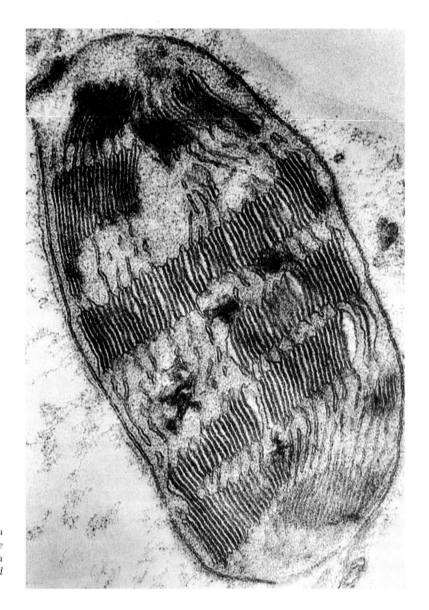

Figure 9-2 **Chloroplasts** (×4900) *from a green plant (Nicotiana rustica). Note the numerous membraneous structures which contain the chlorophyll, other cofactors, and enzymes.* [Courtesy Elliott Weier.]

then exposing them to light until various amounts of chlorophyll are formed. If you use light bulbs of different colors, you can study the effectiveness of different wavelengths of light in making the leaves become green.

There are other pigments in green plants capable of absorbing light and transferring its energy to chlorophyll. Called **accessory pigments,** they are very important in that they allow the plant to use light of different wavelengths for photosynthesis. Were it not for accessory pigments, certain wavelengths of light would not be absorbed and used by the chlorophyll system. The blue pigment, phycocyanin, in blue-green algae is an accessory pigment.

Not only is photosynthesis a supplier of food for living organisms, it is also a supplier of energy for other processes. Our industrial civilization has depended on a reservoir of coal and oil, which we have been removing from the ground at a rapid rate during the past century. Both coal and oil are derived from plants which grew in past geologic ages. Although we cannot say exactly how long these reservoirs will last, at the present we need not worry about what seems to be an inexhaustible supply of fossil fuel. However, the day will come when this stored supply of coal and oil will be used up, and a continuing source of energy will have to be found.

We could, of course, make use of the abundant energy coming from the Sun. It has been calculated, for example, that $1\frac{1}{2}$ square miles of the Earth's surface receives from the Sun during one day approximately the same amount of energy released by the explosion of one small atomic bomb. The difficulty at the present time is that man does not understand how to make really good use of the sunlight directly. About all we can do now is convert sunlight into electricity by using solar batteries. Possibly in the future we may learn how to store this energy in some other more stable form; it could then be released as we needed it.

The need for power sources is one of the reasons for our great interest in photosynthesis. A process which, on a large scale, captures light energy and stores it in a chemical form is of considerable economic interest. Photosynthesis takes place over the surface of our planet and fixes CO_2 at the rate of about a million tons of carbon per minute. Unfortunately, we do not yet know if we can duplicate these feats in the laboratory in an economical way. If we do manage to build a machine capable of converting solar energy to power on a large scale, the machine will probably be quite different from the green plant. Considering that the Earth's energy sources are limited, a photosynthesizing machine, or something akin to it, must one day become a necessity. As one investigator has pointed out, unless more progress is made in solar energy conversion during the next hundred years, we may again go back to the horse and buggy method of transportation.

The visual process is another striking biological example in which excited states must be involved. In spite of a great deal of outstanding work on the structure of the retina and on the biochemistry of the visual pigments in the eye, very little is known about the basic mechanisms underlying vision. By exposing the optic nerve that connects the eye to the brain, researchers have demonstrated that a nerve impulse is produced by light shining on the retina, which indicates that the primary photochemical act must occur in the eye. A man's eye contains about 4 million units called cones and an additional 125 million units called rods; the rods lead to about a million optic-nerve fibers and are responsible for low-intensity vision in the dark, which is predominantly black and white vision. The cones control high-intensity color vision. Because a pigeon's eye contains only cones, pigeons have only day or "bright light" vision; conversely, owls, having only rods, possess night or dim vision.

We assume that the triggering mechanism for the nerve impulse must lie in the rods and cones, because most of the visual pigments are found in the outer segments of those units. One of the visual pigments (rhodopsin) has been extracted and has been shown to consist of a protein (opsin) attached to a derivative of vitamin A (retinene). Vitamin A is converted into retinene by the reduction of the aldehyde group to the alcohol by DPNH. Alcohol dehydrogenase catalyzes this reaction. When light strikes the visual pigments, rhodopsin immediately dissociates into opsin and retinene, and it is during this time that the nerve impulse is triggered. Oddly enough, the retinene that comes off is different from that which goes back on to opsin to make rhodopsin.

The relationship of these retinene isomers to vitamin A is shown in Fig. 9-3. The connection between vitamin A and vision was first noticed when a vitamin A deficiency was observed to affect the eyes' ability to adapt to the dark. When insufficient vitamin A is present, the reconstruction of the visual pigments is retarded and night blindness results. The absorption spectrum of rhodopsin is identical to the sensitivity of the eye to different wavelengths of light. The peak absorption and sensitivity is around 500 nm.

How the photo-excitation of pigment molecules produces an impulse in the fibers is not understood. Early experiments suggested that at least the initiating step was clear. Rhodopsin is known to dissociate when illuminated with visible light and to reform in the dark. The obvious conclusion was drawn from these facts: The photocurrent is

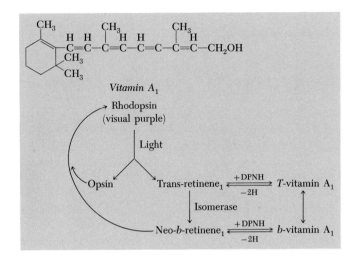

Vitamin A₁ → Rhodopsin (visual purple) → Light → Opsin + Trans-retinene₁

$$\text{Trans-retinene}_1 \underset{-2H}{\overset{+DPNH}{\rightleftharpoons}} T\text{-vitamin A}_1$$

Isomerase

$$\text{Neo-}b\text{-retinene}_1 \underset{-2H}{\overset{+DPNH}{\rightleftharpoons}} b\text{-vitamin A}_1$$

*Figure 9-3 **Some chemical events in vision.***

associated with the changes in retinene. Recent studies, however, have indicated that these changes probably do not constitute the primary event, since the photochemical changes of rhodopsin are quite slow compared to the rate of initiation and conduction of the nerve impulse. It is now believed that the small chemical changes that occur during the visual process are due to some alteration that precedes the transformation in retinene itself. Maybe the photocurrent is caused by some electrolytic process that may result from the SH groups on opsin being exposed when retinene dissociates.

The emission of light by organisms, *bioluminescence*, is a particular example of chemiluminescence. This form of luminescence occurs when the energy released during a biological oxidation forces an electron of a molecule to a higher energy level. In this case, the oxidation energy serves the same purpose as the incident proton does in photoluminescence. That organic chemicals could give off light was first observed in the latter part of the nineteenth century. In 1887 the French physiologist Dubois suggested that the light from a luminous clam, *Pholas dactylus*, is caused by an oxidizable substance he called *luciferin*. Luciferin remains stable when heated, but in the presence of oxygen and an enzyme, *luciferase*, it is destroyed by an oxidative reaction that gives out light. Scientists pursuing this problem have discovered that luciferin is not a single compound common to all

Figure 9-4 **Luminous toadstools.** *Upper: by ordinary light; lower: by own light.* [Courtesy of Dr. Y. Haneda.]

luminous organisms. Apparently, there are many different luciferins, just as there are many different vitamins.

Much work remains to be done, therefore, on the various luminous forms, which range from the bacteria that give off a blue light (495 nm) to the South American railroad worm (larva of a beetle), which gives off a red light (640 nm). The light of decaying wood is produced by fungi (530 nm). The luminescence of the sea is caused by a variety of forms: protozoa, radiolarians, sponges, jellyfish, comb jellies, brittle stars, snails, clams, squid, shrimp, copepods, fish, and many others. Probably the best-known luminous forms on land are the fireflies and glow worms. In addition to these luminous beetles, there are luminous spring-tails, flies, centipedes, millipedes, earthworms, and snails. The only known freshwater luminous animal is the limpit (*Latia*), a native of New Zealand.

The mechanism of bioluminescence is poorly understood. We are confronted with two fundamental, unanswered questions concerning this type of luminescence: (1) What is the nature of the excited molecule? and (2) what is the chemical reaction that is capable of supplying so much energy for the excitation process? As we have seen, in most oxidations of biological interest, the energy is liberated in small units, usually just sufficient to synthesize a pyrophosphate bond (8 kilocalories per mole). Since *visible* radiation is emitted by organisms, much more energy must obviously be liberated by this oxidation process. Although a great deal of progress has been made in recent years in probing the nature of the chemical substances required for light emission, the basic mechanism remains obscure. In some respects, bioluminescence is similar to photosynthesis and vision. In the latter two cases, the excited states are generated by light and chemistry results, while in luminescence the excited state is created by chemical reactions and the energy is lost as light.

A number of investigators have attempted to determine the importance of light emission to those organisms that display this remarkable ability. Unfortunately there is no clear answer as yet. There are many examples in which the bioluminescent reaction has been adapted to good biological use. The reproduction cycles of a number of organisms in the sea are intimately tied to light emission. The flash of the firefly is used by some species as a sex signal. Light emission by deep-sea organisms provides the only light source for those organisms at great depths that have eyes. In some cases luminous bacteria are known to grow in special glands on fish and to provide a regular source of light. Whether light emission in the deeps of the ocean trigger other photobiological processes is not known, but it is a safe guess that it does. A photograph of luminous toadstools is shown in Fig. 9-4.

That there are many different chemical types of bioluminescent reactions suggests that they arose independently during evolution and are an expression of the reverse process of photosynthesis. In photosynthesis oxygen is evolved and organic molecules are produced by various reduction reactions, whereas in bioluminescence oxygen decomposes organic molecules in an oxidative event that leads to an excited state and light emission. It has been suggested that excited states of this type normally occur in oxidative reactions, but the energy is most often used to generate ATP rather than to produce light emission. In fact, one can present a reasonable argument that life arose and evolved by making use of excited states, as plants are capable of doing now in photosynthesis. During the early evolution of life, therefore, many more light-emitting organisms may have existed than we see at the present time. Bioluminescence is perhaps gradually being lost as organisms evolve effective means of utilizing the energy of the excited state. The use of light emission for specific purposes may thus have been secondary in the evolutionary process. Obviously, much work remains to be done in this area of biological energy transformation.

In addition to the photosynthetic reactions in green plants that are dependent on light there are other striking effects of light. Most green plants, for example, display striking growth responses to light, and their response may be independent of photosynthesis. The bending and growth of a green plant toward light is known as **phototropism** and is attributed to the effect of light on the metabolism of one of the plant growth hormones, **auxin.** The duration of light and dark periods also regulates the flowering and reproduction of plants, an effect known as **photoperiodism.** The phenomenon of photoperiodism is not limited to plants. Most organisms that have been studied carefully show a day-night periodicity for a number of physiological processes. Thus both animals and plants exhibit rhythms in a manner indicating that a "biological clock" plays an important role in controlling behavior and physiological function.

Plants and animals are able to "recognize" the seasons of the year by measuring the changing length of night and day. For example, the migration of Canadian geese in the autumn is initiated by the change in the length of the nights and is one of the many important and interesting displays of photoperiodic responses in animals. The length of the day or the shortness of the night have been demonstrated to have important effects in the sexual cycle of numerous fishes, reptiles, and in the reproductive cycle and migration of birds.

The flowering of the poinsettia at the Christmas season is another example of photoperiodism. Other plants grow, flower, and bear fruit at different times of the year. Thus, we come to recognize that some clocks in plants cause flowering in the spring, while other periodic clocks cause flowering in the summer, and still others in the autumn.

Unfortunately, we know very little about the details of the biochemistry or physiology of the biological clock systems that control these rhythmic responses. There must be specific pigments in the cells that are affected by light and, in turn, regulate cellular responses. One such pigment in plants, *phytochrome*, has been studied in detail. We know from studies on flowering response at various wavelengths of light that the interruption of a long dark period by red light inhibits the flowering response in the so-called short-day plants (that is, those that require long night). If the plants are subsequently exposed to far-red light, flowering is stimulated. The pigment which absorbs the light is phytochrome. One form absorbs in the red region and is then converted into a form that absorbs in the far-red region. For flowering to occur in short-day plants (for example, soybean) the period of darkness (or far-red) must be long enough to decrease the concentration of the far-red-absorbing pigment to a low level by conversion to the red-absorbing form. This state must be maintained for several hours if flowering is to be initiated. If the dark period is interrupted with a brief flash of red light and far-red-absorbing pigment produced suppresses flowering.

Unfortunately, we do not know the mechanisms involved in the flowering response. One important lead has been obtained by grafting experiments. By such techniques we can show that the phytochrome is in some way associated with the production of a plant hormone that initiates flowering. Basic research in this area is of considerable importance since it could lead to important agricultural advances.

Flowering in plants is not the only response to the red-far-red light system. Leaf expansion, stem elongation, seed germination, and other processes also demonstrate a red-far-red antagonistic response.

THE PRIMARY CONTROL OF CELL METABOLISM MUST BE ACCOMPLISHED
through the regulation of enzyme synthesis. The constitutive enzymes
of a cell are the main determinants of the cell's form and function.
Although cells have a remarkably wide range of responses to the
environment, cells of the same type show similar properties and similar
responses. For this reason, it is clear that the hereditary mechanisms
of cells are fundamental to the control of cellular activity. Modern
genetics holds that the primary direction of the elaborately interrelated
processes of metabolism, development, and function come from genetic
material carried in the chromosomes of the cell nucleus. The nature
of this material, its method of production, the manner in which it
functions, the mechanisms of its transmission from one generation to
another, and its role in the process of organic evolution are basic factors
in the science of genetics. Although another book° in this series con-
siders the detailed aspects of genetics, we must take up some of these

° D. M. Bonner and S. E. Mills, *Heredity*, 2nd ed. (Englewood Cliffs, N. J.: Prentice-
Hall, 1964).

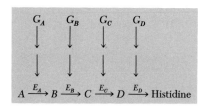

$$G_A \qquad G_B \qquad G_C \qquad G_D$$
$$\downarrow \qquad \downarrow \qquad \downarrow \qquad \downarrow$$
$$\downarrow \qquad \downarrow \qquad \downarrow \qquad \downarrow$$
$$A \xrightarrow{E_A} B \xrightarrow{E_B} C \xrightarrow{E_C} D \xrightarrow{E_D} \text{Histidine}$$

Figure 10-1   **The genetic control of histidine biosynthesis.**

problems here, since the control of cell division, the duplication of the genetic material, and the synthesis of the metabolic units of the cell are important activities in biology that are of primary interest also to physiologists and biochemists.

Geneticists and cytologists have established that the hereditary material in all organisms is basically the same, and careful studies have revealed that this material is transported in the chromosomes. In a particular chromosome, each of the many genes appears to occupy a fixed and special position known as a locus. The order and spacing of the genes in a single chromosome are determined by recombination and linkage tests (which are described elsewhere in this series°). That these positions are correct and are located in the chromosomes can be verified by correlating visible chromosome changes with certain biochemical deficiencies. For example, the absence of a specific small segment of chromosome can sometimes be related to the absence of a specific gene that controls a known biochemical step.

One method of finding out about genes is to study their mutational properties. From such studies, George Beadle and E. L. Tatum concluded that a gene mutation alters cellular processes by affecting protein synthesis. When it was later found that gene mutation changes the nutritional requirement of an organism, it became obvious that the genes control the biosynthesis of the basic chemical units of the cell by controlling the synthesis of specific enzymes that are the essential catalysts for biosynthesis.

For example, Fig. 10-1 presents an outline of a series of hypothetical reactions which lead to the synthesis of the amino acid histidine. Each biochemical step in this synthetic pathway is catalyzed by a specific enzyme (E_A, E_B, etc.). The synthesis of these enzymes, in turn, is under the control of specific genes (G_A, G_B, etc.) in the chromosome. Mutational studies have shown that if we alter gene A we also affect the synthesis of enzyme A and thus prevent the immediate synthesis of B, which is an essential precursor of histidine. If an organism experienced such a mutation, it would require histidine in its diet in order to survive and grow.

Since the hypothesis that genes control enzyme synthesis is reasonably supported by experimental facts, we are confident today that genes exert their control largely through the enzymes they produce. It is now clear that the wide variation that exists in the nutritional requirements of organisms is caused by changes in the genetic constitution of the organisms. The ability of an organism to respond to a particular environment and to gain a specific metabolic ability, therefore, depends on its genetic constitution.

° C. P. Swanson, *The Cell*, 3rd ed. (Englewood Cliffs, N. J.: Prentice-Hall, 1969).

This knowledge is a tremendous step toward our general goal of decoding the inherited message in the nucleus. We are now reasonably certain that the message is chemical in nature and that the genes, through their chemical composition and internal arrangement, control the protein-forming center. The gene's ability to produce an exact copy of itself in each cell generation is its most outstanding property, and this ability could well be explained by its capacity to make an enzyme, thus imparting its character to the enzyme as a mold fashions the shape of a casting. Genetic control of metabolism, therefore, is exercised primarily through the ability of genes to regulate, in some way, the synthesis of proteins with specific primary and secondary structures. We now investigate the possible ways the chemistry of the genetic material may control protein synthesis.

Chromosomes consist mainly of two kinds of chemical substances, protein and deoxyribose nucleic acid (DNA). A second type of nucleic acid, ribose nucleic acid (RNA), also exists in small concentrations in the nucleus, and much larger quantities are found in the cytoplasm. DNA synthesis and chromosome replication are closely connected with the general reproduction process at the subcellular level, and we are confident that the DNA in the chromosome is the chemical substance primarily concerned with the transfer of hereditary information from one generation to the next. Later we shall discuss a number of experiments that support this theory.

When DNA is isolated from almost any source, it is found to consist of thread-like particles of very high molecular or particle weight (several millions). Its chemical composition varies greatly, but these differences are due largely to the arrangement of a few simple chemical compounds called purine and pyrimidine bases. The structure of the two purine bases, adenine and guanine, and the two pyrimidine bases, thymine and cytosine, are shown in Fig. 10-2. In the intact nucleic acid molecule these bases are attached to a five-carbon sugar, 2-deoxyribose, to form a deoxynucleoside. The nucleosides are linked together by a phosphate group at the number 3 position of one nucleoside and the 5 position of a second. When DNA is hydrolyzed by acid, the first products observed are deoxyribonucleotides. Thus we conclude that DNA is a polynucleotide whose chemical backbone is made up of phosphate ester linkages between the sugars. The variation in chemical composition of different DNA molecules must depend, then, on the purine and pyrimidine bases. Present evidence indicates that the sequence of these bases must be the major factor in determining the functional (hereditary) properties of the DNA molecule. Chemical

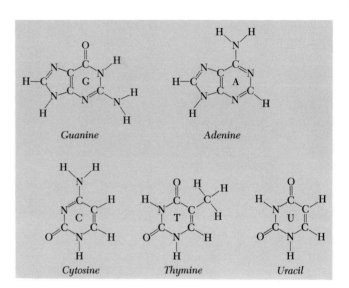

Figure 10-2   ***Components of DNA and RNA.*** *Adenine, guanine, thymine, and cytosine are the four bases that make up DNA. In RNA the thymine is replaced by uracil.*

analyses of a number of samples of DNA indicate that the bases do not occur in a random fashion and that the sum of the purines equals the sum of the pyrimidines; the content of adenine also equals that of thymine and the content of guanine equals that of cytosine. In some DNA molecules, the A plus T content is much larger than the G plus C content, whereas in other preparations the reverse is true. Since the possible combinations of nucleotides in the DNA molecule are of the order of 10^{200}, this molecule has great potential for the transmission of information in the form of a molecular code to future generations. The potentialities of DNA differences are also important when we consider that each species of organism has a different base sequence. The order, however, must be characteristic of the species. Since there are well over two million species of organisms, there must be at least that many different DNA molecules whose identity depends only on base sequence.

With this information about the restrictions on base pairing and with additional physical data on DNA molecules, Watson and Crick in 1953 constructed a three-dimensional model of DNA. Its uniform molecular pattern is produced by two polynucleotide chains arranged in a helical structure, and the two chains are held together by hydrogen bonding between the bases. Chemical data and the symmetry of the helix suggested that adenine-thymine and guanine-cytosine were the base pairs and that the bases on one strand determine the sequence on the partner strand. The hydrogen bonding of these pairs is shown in Figs. 10-3 and 10-4. It is evident from this pairing that if the bases

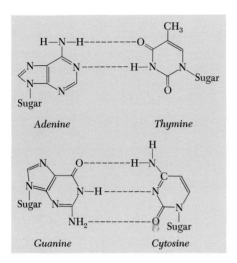

Figure 10-3 *Pairing of purine and pyrimidine bases* by means of hydrogen bonding.

Adenine   Thymine

Guanine   Cytosine

Figure 10-4 *A schematic diagram* of the DNA molecule showing the base pairing and the helical structure.

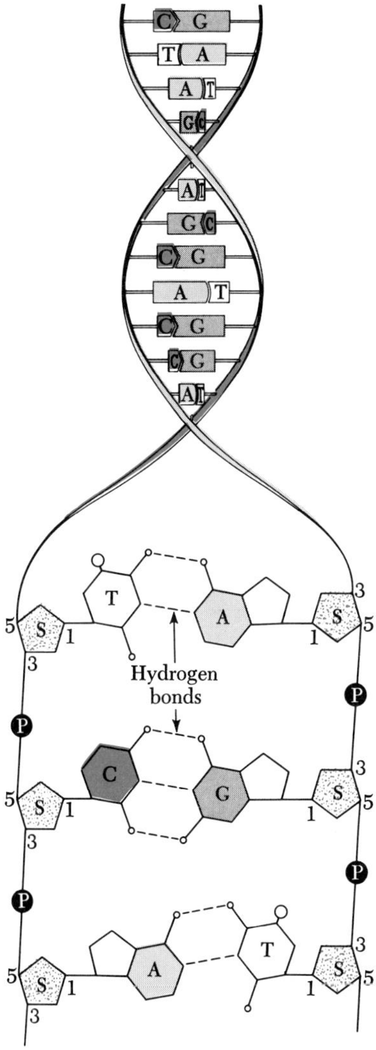

in one strand are arranged in the order GAGGTC, the complementary strand will be CTCCAG (Fig. 10-4). This conception of the DNA molecule is one of the major achievements of biochemistry.

From a biological viewpoint, the proposed DNA structure is quite satisfying, since it is one of the few models so far constructed that reasonably explains how a complex and highly specific molecule can be duplicated from an array of building blocks. With this model, we assume that during the replication process, the complementary poly-nucleotide chains separate and that each one acquires a new partner from the base pool. Experiments using the heavy isotope of N^{15} indicate that when new DNA strands are produced there is no degradation of the old strand, a result that is in keeping with the idea that the old DNA strand acts as a template on which a new strand is then made.

Recent experiments by Kornberg and his colleagues on the enzy-matic synthesis of DNA-like molecules suggest that the base pairing of the new material is determined by the small amount of DNA present initially in the enzyme preparations used. This cell-free synthesis re-quires the presence of the corresponding triphosphates of the purine and pyrimidine bases (i.e., deoxy ATP, TTP, GTP, and CTP), the appropriate enzyme preparation DNA polymerase, and a DNA primer or starter. The enzyme catalyzes the condensation of the deoxyribonu-cleoside triphosphates, liberating pyrophosphate (PP). Interestingly enough, the new DNA contains the same base ratio as the primer, which suggests that the existing DNA acted in some way to determine the chemical sequence.

Hydrogen bonds

CONTROL OF PROTEIN SYNTHESIS

The synthesis of the 5'- mono-, di-, and triphosphates of the purine and pyrimidine bases and the corresponding deoxyribonucleotides involves a number of specific enzyme-catalyzed reactions and will not be covered in this brief text. However, the fundamental principles are not different from those we have discussed so far. For example, in purine synthesis we make use of aspartic acid, CO_2, formic acid, formate, glycine, and glutamine. Pyrimidine synthesis requires the use of carbamyl phosphate, aspartic acid, and an active form of ribose (phosphoribosylpyrophosphate). The corresponding nucleotides thus formed can be converted into di- and triphosphates making use of ATP and specific transphorphylases.

Ribonucleotides diphosphates are the immediate precursors to the deoxynucleotides.

Probably the best direct evidence of DNA's role as a hereditary determinant in metabolism comes from studies of the *transformation* of certain bacteria, particularly pneumococci. These experiments have demonstrated that the DNA from one strain of bacteria can alter the inherited metabolic capacities of a second strain. For example, by mutational techniques a strain of pneumococcus that is incapable of utilizing mannitol as a substrate can be isolated, and these bacterial cells do not contain the enzyme mannitol phosphate dehydrogenase (M^- cells). When DNA is isolated from M^+ cells and added to a culture of M^- cells, many M^- cells are transformed into M^+ cells. The progeny of these cells are capable of making the enzyme, thus indicating quite convincingly that the hereditary capacity of the cells is permanently altered. These results suggest that at least part of the genetic material in DNA is incorporated into the chromosomal structure of the host, and in a certain percentage of the cases this material is replicated during subsequent generations. How this DNA is incorporated into a permanent functional structure we do not know.

A number of experiments involving the transformation of genetic characters have been made, and we are now certain that the transforming agent is DNA. These experiments also strongly support the theory that DNA is the major chemical responsible for controlling the synthesis of the enzymic machinery from one generation to another.

The identification of DNA as the genetic messenger is also supported by new experiments using virus. From recent investigations of bacteriophage (bacterial virus), we know that these parasites attach themselves to the bacterial cell by means of their protein coat and that the DNA of the phage is immediately injected into the cell. Inside the bacterial cell, the phage DNA takes over the metabolic machinery

Figure 10-5   Life cycle of bacteriophage (*T phage of E. coli*).

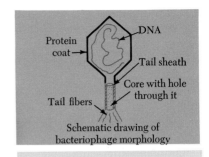

and begins to make phage protein and new phage DNA. Eventually, the protein combines with the DNA to form a completed phage, and all this occurs at the metabolic expense of the host cell. The life cycle of the bacteriophage is schematically presented in Fig. 10-5. The evidence is quite clear that a "foreign" DNA is capable of influencing the synthetic capacities of cells. This fact is largely responsible for our present belief that an altered DNA or a virus DNA may be the cause of a number of diseases, particularly cancer.

Recently, Goulian, Kornberg, and Sinsheimer have demonstrated the cell-free synthesis of a virus that is completely active biologically, i.e., it will infect its natural host, the bacterium *E. coli*. The DNA of this particular virus, ϕX174, is circular; thus, when it acts as a template for DNA polymerase, a second enzyme is required to join the ends of the newly synthesized strand. This second enzyme, DNA ligase, or joining enzyme, uses pyrophosphate bond energy to join the two ends of the polynucleotide strand to make a circle. It may be that DNA ligase is very important during the life cycle of a cell in repairing broken strands of DNA in any part of its structure, whether the DNA is circular or linear.

If one uses a single circular strand of ϕX174 DNA, the two enzymes mentioned above, and the nucleotide triphosphate precursors, it is possible to synthesize in the test tube a complete complementary copy of the viral DNA containing 5500 bases. Either strand is infectious.

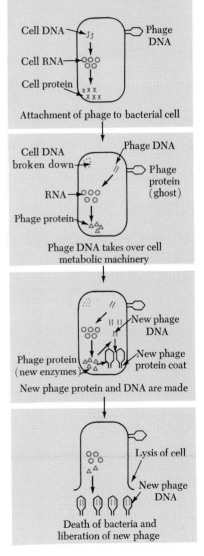

RELATION BETWEEN DNA AND RNA

From what we know about DNA, we can see that it is capable of inducing the formation of specific proteins, although there are numerous cases in which protein synthesis can take place in the absence of DNA. The present evidence indicates that the DNA acts also as the template to form RNA, which then leaves the nucleus and moves into the cytoplasm where it functions in protein synthesis.

There are important chemical differences between DNA and RNA. In RNA the five-carbon sugar attached to the bases, for example, is ribose instead of deoxyribose. The pyrimidine base, thymine, is not found in RNA but is replaced by the base uracil. The four bases that make up RNA, therefore, are adenine, guanine, cystosine and uracil. RNA is also helical in structure and its nucleotides are spaced in much the same way as those in DNA. Recently it has been possible to demonstrate that DNA can act as a template for the synthesis of RNA. The requirements for synthesis are the four triphosphates (ATP, UTP,

GTP and CTP), DNA, and the enzyme RNA polymerase. The bases of RNA can pair to the bases of DNA. Thus, in the two strands of the DNA double helix, guanine and cytosine pair, but when RNA is synthesized on the DNA, the cytosine of the RNA can pair with the guanine of the DNA (and vice versa). Thus, the DNA can act as a template for an RNA "copy" which contains a specific inherited sequence of bases. As we shall discuss later, this specific RNA can be utilized as the template for the synthesis of a specific polypeptide chain.

PROTEIN SYNTHESIS

The problem of how proteins are synthesized by the cell has become more clearly defined with our increasing knowledge of protein structure and the relation of structure to enzymatic activity. But most important of all are the discoveries that indicate that the four bases of DNA make up the "letters" of the genetic code. Various combinations of these bases, or code letters, provide the specific biochemical information used by the cell in the construction of proteins. Apparently each of the 20 amino acids is directed to the proper site in the protein chain by a sequence of code letters; three bases in a sequence appear to be essential for coding one amino acid. If we have only four letters (bases A, T, G, C) in the alphabet to specify a dictionary containing 20 words (amino acids), then we must use at least 3 bases (letters) to code each amino acid. If two bases were used, we would have only 16 code words (4 × 4) but by using 3 base we have 64 (4 × 4 × 4) code words. Thus, if a protein contains 200 amino acids subunits linked in a specific sequence, the gene for controlling the synthesis of this particular protein must contain at least 200 code words or 600 bases in a specific linear arrangement. Since another book* in this series discusses the details of coding, we shall concern ourselves at this time only with the actual mechanism of protein-synthesizing machinery which permits the code in a nucleic acid to be read and translated for the accurate positioning of the amino acids in a polypeptide chain.

AMINO ACID ACTIVATION   The first step in protein synthesis is the selection of specific amino acids out of the heterogeneous mixture of metabolites in the cytoplasm of the cell. This selection process is carried out by amino-acid-activating enzymes that result in the formation of amino acid adenylates (see Fig. 10-6). The activation of the amino acids requires ATP energy and the reaction is very similar to the activation of fatty acids, which we discussed previously. There is

*D. M. Bonner, *Heredity*, 2nd ed. (Englewood Cliffs, N. J.: Prentice-Hall, 1964).

Figure 10-6   The activation of amino acids for protein synthesis.

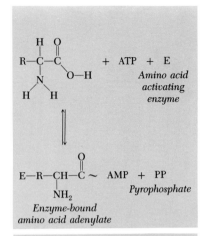

at least one unique activating enzyme for each different amino acid incorporated into protein, and the specificity of this enzyme for the particular amino acid is very high, insuring critical control of protein synthesis.

tRNA—THE ADAPTOR   The activated amino acids remain tightly bound on the enzyme following the activation step. The next step in protein synthesis involves the attachment of the amino acid to a specific adaptor RNA molecule. This RNA, a small, soluble molecule, has been variously termed soluble (s), transfer (t) or adaptor RNA. As we shall discuss shortly, the existence of several classes of tRNA to which a single amino acid is attached is important in the coding process.

tRNA is a single chain of about 70 nucleotides that is folded back upon itself and held in a helical configuration by hydrogen bonding. Consequently the molecular weight is about 25,000. The terminal sequence of nucleotides to which the amino acid is attached has been shown to be in all cases adenylic-cytidylic-cytidylic (A-C-C). The amino acid-adenylic acid from the activating reaction interacts with tRNA and the amino acid is transferred to the second or third carbon of the ribose molecule in the terminal adenylic acid of tRNA (see Fig. 10-6).

Although all tRNA appears to be similar in gross structure, there is much indirect evidence which indicates some structural characteristics that are unique to each class and specific for a given species of activating enzyme. As discussed later, the tRNA molecules must contain a special short sequence, the anti-codon, that allows them to seek out the proper complementary series of bases in the special RNA (messenger) found at the protein-forming site (see Fig. 10-8).

RIBOSOMES   The protein-forming system uses the amino acid-tRNA complexes as the raw material for making a polypeptide chain. The biochemical events in protein synthesis take place on small granules (ribosomes) that are located in the cytoplasm. Ribosomes contain about 60 per cent RNA and 40 per cent protein. In bacteria they are found throughout the cytoplasm whereas in higher organisms they are often found attached to the surface of tubular lipoprotein membranes found in the cytoplasm. This membrane system is called collectively the endoplasmic reticulum (see Fig. 10-7). Ribosomes have been shown to be the site for protein synthesis in a number of indirect ways; however, much stronger support has been obtained recently by the isolation of the ribosomal particles free from other cellular components. Such isolated ribosomes, under appropriate conditions, can incorporate

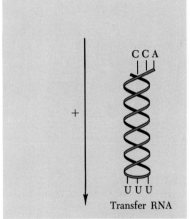

Transfer RNA

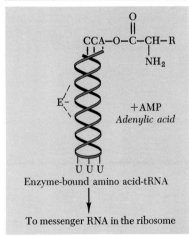

Enzyme-bound amino acid-tRNA

To messenger RNA in the ribosome

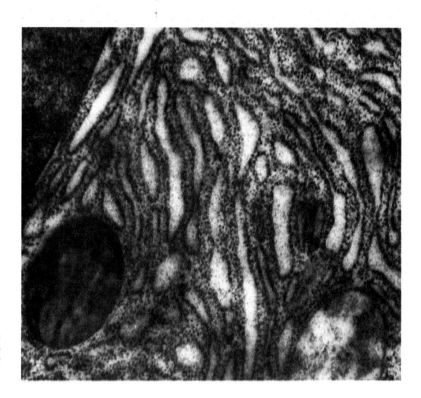

Figure 10-7 **Endoplasmic reticulum** *in a pancreatic cell, showing ribosome particles and a large zymogen granule.* [Courtesy of George Palade.]

amino acids from the t-RNA-AA complex into a polypeptide chain. Ribosomes are rather complicated structures which can dissociate into smaller inactive subunits. The function of the structural RNA of the ribosomes is not known, although it seems reasonably clear that it is not directly involved in coding of the sequence of amino acids into protein.

mRNA—THE MESSENGER   Research during the past few years indicates that a special class of RNA molecules acts as the template upon which proteins are built. These are called messenger RNA (mRNA) since, as recent evidence indicates, they are ribopolynucleotides that are direct copies of the genetic messages from DNA.

A number of workers have shown that an enzyme, RNA polymerase, catalyzes the synthesis of mRNA and that DNA is required as the primer (see Fig. 10-8). Additional evidence indicates that the enzyme catalyzes a preferential copy of the base sequence in DNA rather than that it joins nucleotides in a random fashion. Not all messenger RNA is made with DNA as the template. As indicated previously, certain viruses contain RNA instead of DNA as the genetic material and it seems likely that this RNA can act directly as messenger for the synthesis of virus protein.

After mRNA is synthesized it leaves the template and is adsorbed onto the ribosome. This mRNA-ribosome complex is the active unit for protein synthesis. If we add to this complex the 20 amino acids, the various tRNA's, ATP, and the activating enzymes, we observe the synthesis of a polypeptide chain. In the formation of the peptide, additional enzymes are required as well as guanosine triphosphate (GTP).

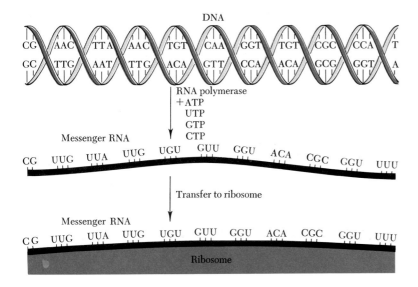

Figure 10-8   A scheme illustrating the role of DNA and RNA in protein synthesis.

There are numerous experiments which indicate that the mRNA is the specific template for the synthesis of specific polypeptides and that activating enzymes and tRNA are used over and over again in the synthesis of a number of different proteins. Probably the most striking example of this is furnished by the experiments of Nirenberg in which he used synthetic messengers. Synthetic ribopolynucleotides can be made by using an enzyme, polynucleotide phosphorylase (which probably serves normally to degrade RNA), and the appropriate nucleotide triphosphates. The composition of the product depends almost entirely on the concentration of the nucleotides. It has been possible, for example, to synthesize a polymer containing only uracil (polyuridylic acid or poly U). When poly U was added to a ribosomal system along with the other components essential for protein synthesis, it was found that only phenylalanine was incorporated into the polypeptide. Thus it would appear that the triplet code for this amino acid was UUU (see Fig. 10-9). This system has proven very useful in translating the genetic code for other amino acids. Poly C (CCC) appears to be the code for the amino acid proline, CCG for alanine, GUA for aspartic acid, ACC for histidine, and so on.

The genetic code for all of the amino acids is shown in Table 10-1. It is clear that more than one codon can code for a single amino acid. For example, UUU as well as UUC can code for phenylalanine. This so-called degeneracy seems to be essential, for if we use only two bases, there would be only 16 (4^2) possible sequences for coding the 20 amino acids. The three-letter code gives 64 (4^3) sequences for 20 amino acids, thus allowing for some degeneracy.

Thus the specific organization and alignment of amino acids in the polypeptide chain of a protein is dictated by the base sequence in the

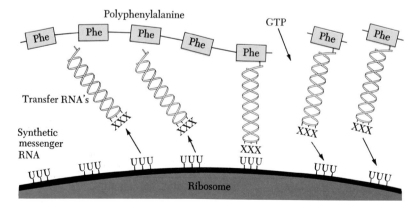

Figure 10-9 *The first break in the genetic code was the discovery that a synthetic messenger RNA containing only uracil (poly-U) directed the manufacture of a synthetic protein containing only one amino acid, phenylalanine (PHE). The X's in transfer RNA signify that the bases that respond to code words in messenger RNA are not known.* [After Nirenberg.]

Table 10-1   *Genetic code for amino acids*

FIRST POSITION	SECOND POSITION				THIRD POSITION
	U	C	A	G	
U	PHE	SER	TYR	CYS	U
	PHE	SER	TYR	CYS	C
	LEU	SER	OCHRE	UMBER	A
	LEU	SER	AMBER	TRP	G
C	LEU	PRO	HIS	ARG	U
	LEU	PRO	HIS	ARG	C
	LEU	PRO	GLN	ARG	A
	LEU	PRO	GLN	ARG	G
A	ILE	THR	ASN	SER	U
	ILE	THR	ASN	SER	C
	ILE	THR	LYS	ARG	A
	MET°	THR	LYS	ARG	G
G	VAL	ALA	ASP	GLY	U
	VAL	ALA	ASP	GLY	C
	VAL	ALA	GLU	GLY	A
	VAL°	ALA	GLU	GLY	G

° *Chain initiating codon.*

mRNA which in turn is a direct reflection of the base sequence in the gene (DNA). The specific adaptor molecule, tRNA, is responsible for carrying a specific amino acid to the site for protein synthesis. The growth of the polypeptide chain (i.e., peptide bond formation) requires GTP and specific enzymes. When the polypeptide is complete it is released from the ribosome and becomes folded in the active configuration. Much remains to be discovered about this latter process and indeed many important questions still remain to be answered in regard to this general scheme of cell-free protein synthesis. Even so, this model has been very useful in explaining a number of genetic, or heritable, changes (mutations). A mutation appears to be due to a specific change in one of the bases in DNA, and this seems to be eventually reflected in the mRNA and finally the protein. One of the best-known examples of this takes place in the hemoglobin of a person with sickle-cell anemia. In this disease, which is inherited as a single gene difference, the hemoglobin is electrophoretically different from the normal. Analysis of the amino acid sequence of the altered hemoglobin reveals that one particular glutamic acid residue is replaced with valine, and this alteration is sufficient to account for the change in charge. A few examples of hereditary disorders are listed in Table 10-2.

Table 10-2   A few examples of hereditary disorders in man (after White, Handler, and Smith)

DISORDER	AFFECTED ENZYME
Albinism	Tyrosinase
Fructose intolerance	Fructose-1-phosphate adolase
Fructosuria	Fructokinase
Goiter (familial)	Iodotyrosine dehalogenase
Hemolytic anemia	Pyruvate kinase
Histidinemia	Histidase
Hypophosphatasia	Alkaline phosphatase
Maple syrup urine disease	Amino acid decarboxylase
Phenylketonuria (PKU)	Phenylalanine hydroxylase
Xanthinuria	Xanthine oxidase

There are some unusual problems in connection with the initiation of the synthesis of a polypeptide chain. In *E. coli* it has been observed that the most frequent N-terminal amino acid found in proteins is methionine. In isolated systems N-formyl methionine is the actual initiator of polypeptide chain formation and if this occurs *in vivo* the formyl group is removed after the chain is released from the ribosome. As indicated in Table 10-1, the codon for methionine is AUG. However, there are two known tRNA's specific for methionine. When methionine is attached to one of these tRNA molecules it can be rapidly formylated at the amino group. Blocking the amino group prevents it from reacting with carboxyl groups of other amino acids to form a peptide bond. Thus N-formyl methionine cannot be incorporated at internal positions in the polypeptide but could function to initiate chain formation.

N-formyl methionine does not seem to play this same role in animals or higher plants. Acetylserine may function in this capacity for some systems in animals. Much remains to be done before we completely understand the process of protein synthesis.

As indicated in Table 10-1, there are three codons, UAA (ochre), UAG (amber), and UGA (umber) that do not code amino acid incorporation into a polypeptide chain. Recent evidence indicates that these are chain-terminating codons. The biochemical mechanism of this process is not clear. It may be a simple passive process in which there is no tRNA for these codons. Therefore, when the growing polypeptide chain reaches one of these triplets, peptide bond formation ceases and the attached polypeptide is hydrolyzed from the last tRNA used.

Other aspects concerning the control of protein synthesis will be considered in the next chapter.

IN ORDER TO FUNCTION, A LIVING CELL MUST RELY ON A COMPLEX SERIES
of reactions. The glycolytic and oxidative cycles, and fatty acid and
amino acid metabolism, are examples of interdependent activities that
affect all cells. In addition, cells that have specific functions must
channel their major energy expenditures through special pathways that
enable them to perform their particular tasks. To regulate and to
coordinate the multi-enzyme systems of the cell so they will achieve
a specific final result require precise control mechanisms. One of the
greatest challenges facing the cell physiologist today is that of piecing
together our knowledge of the active framework of the living cell, the
framework that supports the ordered processes of life.

From our earlier considerations of cell metabolism, it is evident
that both the rate of reactions and the direction they will take is
dependent on the relative concentrations of substrate and enzyme
molecules. We must consider the cell as a vast steady state system,
in which control can be achieved only through alterations in the
concentrations of the reactants. In a very crude extract of yeast cells,
which will ferment carbohydrate in a reasonably ordered manner, the

rate of carbon dioxide production (and of alcohol synthesis) can be controlled by the alteration of the concentrations of such reactants as sugar, phosphate, ATP, DPN, etc. Obviously, within the intact cell, the same factors will exert controlling influences. For example, when a muscle cell contracts suddenly, ATP is broken down, forming inorganic phosphate and ADP, both of which are essential in the stimulation of carbohydrate breakdown. Inorganic phosphate stimulates the phosphorolysis of glycogen to form glucose-1-phosphate, while ADP tends to remove phosphorylated compounds in the triose phosphate dehydrogenase reaction. In this way, the muscle cell can "know" when it needs energy for contraction and relaxation and can supply that need.

In addition to such substrate effects, evidence accumulated over the past few years indicates that the reactants (substrates, enzymes, etc.) do not mix as readily in the cell as they do in the test tube. The reason for this is that, within the cell, enzymes may be bound into the various internal structures so that enzymes and substrates are spatially separated. We have already seen that the mitochondria are cellular units which contain the enzymes concerned with the oxidative metabolism of carbohydrates, as well as with the metabolism of fatty acids and amino acids. In several cases that have been closely studied, the enzymes that are functionally related are apparently tightly bound together within the structure of the mitochondrion. Thus, the enzymes of the Krebs citric acid cycle and those concerned with oxidative phosphorylation seemingly need to be in close proximity for proper functioning. On the other hand, the mitochondrial membrane is not permeable to many substrates, which insures a separation between the oxidative and nonoxidative metabolism of carbohydrates.

Many substances of small molecular weight, such as ATP, inorganic phosphate, coenzyme A, and acetylcholine, are also tightly bound and highly localized in cells. Since enzyme systems and their cofactors which are fixed to insoluble units of the cell may be of primary significance in the regulation of cellular activities, we cannot limit our discussion of the control of cell metabolism to the enzyme activity we view in the test tube. We must also examine the structural organization of enzyme systems—in other words, the cellular architecture. However, the localization and separation can only affect the relative concentrations of reactants, so we are safe in saying that variations in metabolic activity caused by changes in this structural organization result from differences in either the enzyme concentration or enzyme activity.

It must be obvious from this discussion that those compounds that are in the lowest concentration and in the greatest demand will be the ones that control the metabolism with the greatest sensitivity. The low concentrations of cofactors such as ATP and DPN provide a method for controlling both the rate and the direction of the entire metabolism, although the depletion of even these factors takes more time than is required for a cell to respond to certain environmental stimuli. We can thus term such mechanisms "slow-responding" controls.

In addition to altering the availability of substrates and cofactors, reactions can be controlled by changing the total amount of active enzyme (as contrasted to the distribution of enzyme). When we compare the enzyme activity of cells under different physiological states, we must determine whether observed increases or decreases are caused by alterations in the activity of pre-existing enzyme molecules or by the synthesis of new molecules. The time interval involved is crucial in such studies. A change that occurs in a short time is of great diagnostic value, since it most likely indicates enzyme activation or inhibition. To explain the very fast off-and-on effects produced in cells by stimulation, we must go beyond the simple fact that the enzyme and substrate are isolated from one another. Minor structural alterations, which convert an enzyme inhibitor into an active form or vice versa, may be responsible for the faster mechanism.

At the present time, we know of many instances in which enzymatic activity can be demonstrated in a cell-free extract only after an inhibitor has been removed. Some of these inhibitors are actually proteins themselves and can be inactivated by heat. The products of an enzyme-catalyzed reaction can also act as inhibitors if they are tightly bound to the enzyme; in this case, the enzyme is enzymatically inactive until the product is removed. In addition the product may regulate a metabolic pathway by inhibiting an enzyme concerned with an earlier step. This mechanism of control, called *feedback inhibition*, may be of considerable importance in regulating the amount of a particular product formed in a biosynthetic pathway. For example, in the biosynthesis of histidine by the bacterium *E.coli* it has been found that there are at least ten different steps involved requiring eight different enzymes. If one adds histidine to the culture where the cells are growing, they stop making this particular amino acid and use the exogenous source of histidine for protein synthesis. This blockage of histidine synthesis by the cell is due to the inhibition by histidine of the enzyme which catalyzes the first step in the biosynthetic pathway. Thus, by this mechanism of feedback inhibition the cell's machinery and energy supply is relieved of an additional duty. Since the first step in the biosynthetic pathway is inhibited, wasteful intermediates are not accumulated.

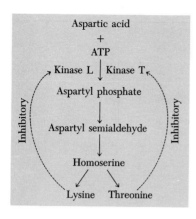

Aspartic acid
+
ATP

Kinase L | Kinase T

Aspartyl phosphate

Aspartyl semialdehyde

Homoserine

Lysine   Threonine

Inhibitory   Inhibitory

Figure 11-1   Regulation of the biosynthesis of lysine and threonine by feedback inhibition.

Feedback inhibition is of particular interest when two essential products are formed from a common intermediate. The synthesis of lysine and threonine is an excellent example. The pathway is briefly outlined in Fig. 11-1. The phosphorylation of aspartic acid to form aspartyl phosphate is the first step in the biosynthesis of these two amino acids. It has been observed that both lysine and threonine independently will partially inhibit the first reaction. The limited inhibition occurs because there are two distinct kinases which catalyze this first step; one is completely inhibited by lysine but not by threonine, the other by threonine but not by lysine. Recent evidence indicates that there are specific sites on the enzymes where the amino acids may combine and thus alter the enzyme structure in such a way that it is no longer catalytically active. This is a significant modification of feedback inhibition, for it is clear that if we had only one aspartic acid kinase, excess lysine would lead to a nutritional requirement for threonine and vice versa. With two kinases that have different properties the cell can effectively regulate the synthesis of these two amino acids.

This general feature of regulation of enzyme activity by feedback has been called *allosterism* by Monod and Jacob because the site on the enzyme molecule where a stimulator or inhibitor molecule might combine is a different site from the substrate combining site. Thus the allosteric site is one that regulates the activity of the enzyme. This stimulation or inhibition depends on the dissociation of the enzyme into subunits or in a change in shape of the enzyme molecule.

ENZYME SYNTHESIS:
INDUCTION AND REPRESSION

There is another way in which the substrate or product of a reaction can affect enzymatic activity of a cell or tissue. This is due to the influence of these substances on protein synthesis. It has been known for some time that the amount of enzyme in cells can be altered by the nutritional or chemical environment in which the organisms are grown. Careful studies of organisms that have been grown under various conditions indicate that the enzymatic composition of the cells depends on the nutrients in the environment. The ability of an organism to respond to a particular environment and to gain a specific metabolic activity depends also on its genetic constitution. However, simply because the organisms have the genetic constitution to make a specific protein does not always mean that the particular protein or enzyme will be present. For example, certain strains of *E. coli* cells, when grown on glucose and ammonia as the primary carbon and nitrogen sources, are capable of making a number of enzymes that are not present in these cells. One such enzyme is beta-galactosidase which catalyzes the hydrolysis of beta-galactosides. A typical substrate is the disaccharide lactose, which occurs in milk. Lactose contains the two hexoses, glucose and galactose, which can be readily metabolized by the *E. coli* cells. If one adds lactose to cells that have been grown on glucose, however, there is a long lag period before the cells can metabolize it. After a time, though, the cells acquire the ability to metabolize lactose rapidly, and a number of experiments indicate that during this lag period the enzyme beta-galactosidase is synthesized by the cells. The results indicate that specific substrates in the environment can induce the formation of new and specific enzymes that are essential for their metabolism. Such enzymes are called *inducible (adaptive)* in contrast to the *constitutive* enzymes that are normally present in the cells when they are grown on glucose. The rate of synthesis of the enzyme in growing *E. coli* is proportional, over a limited range, to the amount of the substrate or inducer that is added. The induction is quite specific since only certain specific beta-galactosides are capable of inducing enzyme formation. There are some inducers that can elicit enzyme formation yet are not metabolized. Such inducers have been used to study the mechanism of the induction process.

Recently it has been possible to demonstrate the formation of specific messenger RNA's during the induction process. This raises the question of how the substrate can induce the synthesis of a specific messenger RNA. The present hypothesis concerning the mechanism of induction stems from observations that by genetic changes it is possible to convert what was normally an inducible enzyme system into a constitutive one. That is, added inducers are not essential for the formation of the messenger RNA and consequently the synthesis of the enzyme. This fact gave rise to the idea that there are both *structural* genes, which are concerned directly with the synthesis of mRNA and consequently the structure of protein, and *regulatory* genes, which are concerned with turning on and off the function of the structural genes. It has been suggested further that the regulatory genes produce a *repressor* substance that can inhibit the action of the structural gene. The inducer apparently combines either with the regulatory gene to prevent the formation of repressor substance or with the repressor itself, which would allow specific mRNA formation.

This brings us again to consideration of the product inhibition phenomena we discussed previously and to the observation that enzyme formation can be prevented by the products of enzyme action. One system that has been studied in great detail concerns the pathway of arginine biosynthesis and the repression of the enzyme ornithine trans-carbamylase by arginine. These relationships are shown in Fig. 11-2

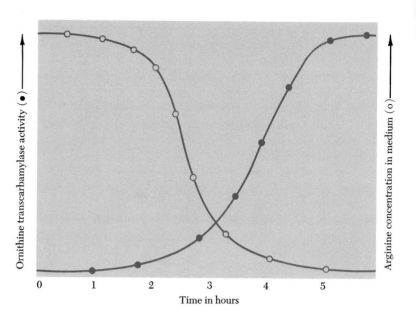

Figure 11-2   Control of enzyme synthesis by substrates (repression).

Ornithine transcarbamylase activity (•)

Arginine concentration in medium (o)

Time in hours

(see also Fig. 8-11 on page 101 concerning the relationship of ornithine to arginine). When cell growth is initiated in a medium containing a rather high arginine concentration it is found that the enzyme (ornithine transcarbamylase) activity is quite low. During growth, however, the organisms use arginine for protein synthesis and consequently the concentration in the medium drops. As the arginine concentration in the medium drops, the activity of ornithine transcarbamylase increases. Careful analysis of this process indicates that the increase in activity is due to the increased rate of synthesis of ornithine transcarbamylase as the arginine concentration goes down. Thus we say that arginine represses the formation of ornithine transcarbamylase. This is different from the feedback inhibition in that not only is the biosynthetic pathway blocked but actually the cell has not made the necessary enzyme for one of the steps in arginine biosynthesis.

It is evident from this example and our discussion of the formation of enzymes, that induction and repression of enzyme synthesis are similar phenomena. It now appears that enzyme induction can be described or interpreted as a *derepression* of an enzyme-forming mechanism of an organism. Genetic changes can release the repression, which makes induction unnecessary. All of these ideas and results, which are still far from clear, suggest that repression must result by some substance or substances that combine with the genetic site essential for the synthesis of a specific protein or enzyme. The inducer may prevent the formation of the active repressor or it may actually combine with and inactivate the repressor, thus allowing the gene for the particular protein to be active. Recently it has been possible to demonstrate that once the inducer has stimulated the formation of messenger RNA the inducer is no longer necessary for enzyme synthesis; this suggests strongly that the repressor and inducer must be working on the DNA in the nucleus.

Feedback inhibition, repression, and induction are important phenomena in the control and regulation of cell metabolism. This type of control must have been of considerable significance during the evolution of organisms. Enzymes would be formed by organisms for the metabolism of a particular substrate only when that substrate is present. The real advantage in being able to make specific enzymes only under certain conditions is selective, because of the more efficient use of energy and metabolites for the synthesis of the parts of cells. Inducibility therefore provides a selective advantage over those cells that are incapable of making a particular enzyme. In addition, as a consequence of this regulated type of enzyme synthesis, cells are prevented from making excess concentrations of enzymes that may in themselves be toxic to the cell.

The regulation of enzyme or protein function can be altered in still another way, one that depends almost entirely on changes in molecular configuration of the protein itself. For example, it has been shown that the enzyme lactic acid dehydrogenase is not a single polypeptide chain that is folded into a secondary structure; rather it is composed of four associated polypeptide chains. Thus, lactic acid dehydrogenase is a tetramer that can be dissociated into two distinct varieties of monomers. These two different types of monomers have been referred to as monomer A and monomer B. Since there are a total of four monomers in one enzyme molecule, it has been suggested that five different types of lactic acid dehydrogenases should be found under appropriate conditions. These would be as follows: LDH_5 would contain four A monomers and no B monomers, i.e., A^4B^0, $LDH_4 = A^3B^1$; $LDH_3 = A^2B^2$; $LDH_2 = A^1B^3$; $LDH_1 = A^0B^4$. Recently it has been possible to purify the LDH_5, which contains only A monomers, and LDH_1, which contains only B monomers. By mixing together in the test tube, dissociating them with salt into monomers, and allowing them to reassociate, it has been possible to demonstrate that the five basic lactic acid dehydrogenases are reformed from the corresponding monomers. This is

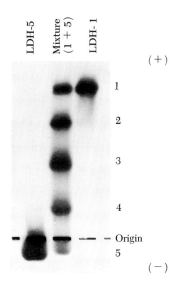

LDH-5 Mixture (1 + 5) LDH-1

(+)

1

2

3

4

Origin

5

(−)

Figure 11-3 **Isozymes of lactic acid dehydrogenase.** [Courtesy of Dr. C. L. Markert.]

illustrated in Fig. 11-3 in which the various lactic acid dehydrogenase enzymes have been separated on a starch electrophoresis strip. The enzymatic activity is indicated by the dark bands. During electrophoresis it is clear that the B monomer—that is, the one that makes up LDH_1—is more negative in charge and therefore moved to the positive electrode, whereas the A monomer of LDH_5, which is positively charged, moves to the negative pole. The various other LDH's have an intermediate charge. These various LDH's, which contain different numbers of subunits of the two types, are called *isozymes*. The functional significance of these different isozymes has recently been investigated, and it has been found that the LDH_1 occurs predominantly in the heart and certain other muscle tissues whereas LDH_5 occurs predominantly in skeletal muscles. Thus, LDH_5 is prevalent in tissues that function under conditions where the concentration of oxygen is low and lactic acid accumulates.

The isozyme composition of tissues changes during embryonic development. In addition, certain diseases and treatments with hormones will alter the isozyme pattern. These induced changes in the LDH composition are in accord with the apparent functional significance of the different isozymes. They demonstrate also that differential synthesis of related polypeptides can be induced in specific tissues by specific agents. Whether these agents act directly at the gene level or at some later point in the sequence from gene to protein is not yet known. Examples of subunits of proteins are given in Table 11-1.

A somewhat similar phenomena that has recently been described demonstrates that by confirmational changes in a protein it is possible to change enzymatic activity not only qualitatively but quantitatively. For example, the enzyme glutamic acid dehydrogenase from mammalian livers is normally made up of four subunits. When these are dissociated into corresponding monomers they no longer have glutamic acid dehydrogenase activity but rather have an alanine dehydrogenase activity. In addition, the state of aggregation of the monomers is subject to regulation by steroid hormones. This fact suggests that some hormones may act by controlling polymerization of polypeptides and by this means regulate metabolic activity. Such mechanisms permit a very rapid reversible shift in function without involving *de novo* protein synthesis. In addition to the steroid hormones, it has been found that DPNH and guanosine triphosphate promote disaggregation whereas ADP promotes aggregation. The availability of these small molecules is closely tied to cell metabolism and their interaction on protein polymerization bridges the gap between the metabolism of small molecules and the regulation of the activity of specific macromolecules.

Table 11-1 *Examples of subunits of proteins (after Klotz and Darnall)*

| | MOLECULAR | SUBUNITS | |
PROTEIN	WEIGHT	NUMBER	MOLECULAR WEIGHT
Insulin	11,466	2	5,733
Bovine growth hormone	48,000	2	25,000
Hemoglobin	64,500	4	16,000
Glycerol-1-phosphate dehydrogenase	78,000	2	40,000
Liver alcohol dehydrogenase	80,000	4	20,000
Enolase	82,000	2	41,000
Tyrosinase	128,000	4	32,000
Fructose diphosphatase	130,000	2	29,000
		2	37,000
Histidine decarboxylase	190,000	10	19,000
Glucose-6-phosphate dehydrogenase	240,000	6	43,000
Acetoacetate decarboxylase	340,000	6	62,000
Phosphorylase A	370,000	4	92,500
β-galactosidase	520,000	4	130,000
Arginine decarboxylase	850,000	5	165,000
Turnip-yellow mosaic virus	5,000,000	150	21,000
Tobacco mosaic virus	40,000,000	2130	17,500

It is becoming increasingly clear that the mechanisms controlling the synthesis of specific proteins and their association or dissociation into units of different functional activity may be of extreme importance during the differentiation of cells and in fact may be fundamentally responsible for differentiation.

Hormonal control of enzymatic activity is not always due to association or dissociation of proteins subunits, however. For example, the hormone adrenalin (epinephrine) apparently works by stimulating an enzyme activity leading to the formation of a cofactor essential for the activation of the enzyme phosphorylase, which is required for the breakdown of glycogen. Thus, when the adrenal glands are stimulated to release adrenalin, the adrenalin in turn stimulates the breakdown of stored glycogen by promoting phosphorylase activity. For the details of other types of functions of hormones that are known, you are referred to advance texts.

We have discussed previously the general problem of cellular permeability and the active transport of nutrients across the cell membrane. Recently it has been shown that the catalytic units in the cell membrane which are capable of mobilizing cellular energy for active accumulation of substrates are under genetic control. For example, we have known for some time that certain bacterial cells contain enzymes for the metabolism of specific compounds but that the cells are unable to use them because the compounds cannot pass the cell membrane. By genetic changes it has been possible to obtain specific mutants whose membrane activity has been altered in some way so that the cells become able to accumulate the substrates. This change allows the cell to metabolize the substrates rapidly. These active transport systems in the membrane, called *permeases,* respond to environmental and genetic changes in much the same way as enzyme synthesis does. It seems definite at the present time, however, that these permeation systems, which are functionally specialized, are distinct from the metabolic enzymes found in the cell. At least in bacteria it seems likely that the entry of most of the organic nutrients a cell metabolizes is mediated by very specific permeation systems. Even in higher forms such as the vertebrates there is considerable evidence for the existence of carriers within membranes and many attempts are being made to isolate and characterize these carriers at the present time. One of the systems that have been studied in great detail involves the differential movement of sodium and potassium ions. Since the transport of sodium and potassium is a process deriving its energy from ATP, it has been suggested that an ATPase may be the transporting enzyme. In fact, some evidence using isolated membranes has been obtained that supports this idea.

The specific permeases present in the cell membrane provide a number of regulatory mechanisms for growth, substrate utilization, enzyme induction or repression, and even feedback inhibition of enzymes. Unfortunately, although we can describe in some detail active transport as well as specific carriers for specific molecules, the mechanism at the molecular level for any of these processes has not been clearly analyzed. But recent studies concerning the transport of ions across membranes suggest that a phospholipid may be important in the transport process as well as ATP energy. It is clear, however, from the standpoint of regulation of growth, that the chemical composition of a cell interior can be regulated by changes in the catalytic character of the cell membrane. Changes in permease activity may be another crucial factor in the regulation of growth, differentiation, and cellular function. Function of other membranes within the cell and their ability to actively mobilize and transport nutrients may also be of some significance in the control of cell metabolism.

REFERENCES

TEXTBOOKS OF BIOCHEMISTRY

KARLSON, P. *Introduction to Modern Biochemistry*. 2nd ed. New York: Academic Press, Inc., 1967.

MAHLER, H. R., AND E. H. CORDES *Biological Chemistry*. New York: Harper & Row, Publishers, 1966.

WHITE, A., P. HANDLER, AND E. SMITH *Principles of Biochemistry*. 4th ed. New York: McGraw-Hill Book Company, 1968.

TEXTBOOKS OF PHYSIOLOGY

DAVSON, HUGH *A Textbook of General Physiology*. 2nd ed. Boston: Little, Brown and Company, 1959.

GIESE, A. C. *Cell Physiology*. 2nd ed. Philadelphia: W. B. Saunders Co., 1963.

SELECTED READINGS

BAILEY, J. L. *Techniques of Protein Chemistry*. 2nd ed. New York: American Elsevier Publishing Co., Inc., 1967.

BALDWIN, E. *Dynamic Aspects of Biochemistry*. 3rd ed. New York: Cambridge University Press, 1957.

DAVIDSON, E. A.   *Carbohydrate Chemistry.* New York: Holt, Rinehart & Winston, Inc., 1967.

DAVIDSON, J. N.   *The Biochemistry of Nucleic Acids.* 5th ed. New York: John Wiley & Sons, Inc., 1965.

DEVEL, H. J., JR.   *The Lipids: Their Chemistry and Biochemistry.* Vols. I, II, and III. New York: Interscience Publishers, Inc., 1951, 1955, 1957.

DIXON, M., AND E. C. WEBB   *Enzymes.* 2nd ed. New York: Academic Press, Inc., 1963.

HEFTMANN, E., ED.   *Chromatography.* 2nd ed. New York: Reinhold Publishing Corp., 1967.

LEHNINGER, A. L.   *The Mitochondrion: Structure and Function.* New York: W. A. Benjamin, Inc., 1964.

————.   *Bioenergetics.* New York: W. A. Benjamin, Inc., 1965.

NYHAN, W. L., ED.   *Amino Acid Metabolism and Genetic Variation.* New York: McGraw-Hill Book Company, 1967.

SCHERAGA, H. A.   *Protein Structure.* New York: Academic Press, Inc., 1963.

SINGER, T. P.   *Biological Oxidations.* New York: John Wiley & Sons, Inc., 1967.

STANBURY, J. B., J. B. WYNGAARDEN, AND D. S. FREDRICKSON.   *The Metabolic Basis of Inherited Disease.* 2nd ed. New York: McGraw-Hill Book Company, 1966.

WATSON, J. B.   *Molecular Biology of the Gene.* New York: W. A. Benjamin, Inc., 1965.

INDEX